Xiaohui Tang

Separating arsenic oxyanions from natural waters for oxygen isotope analysis

Karlsruher Mineralogische und Geochemische Hefte
Schriftenreihe des Instituts für Mineralogie und Geochemie,
Karlsruher Institut für Technologie (KIT)
Band 40

Separating arsenic oxyanions from natural waters for oxygen isotope analysis

by

Xiaohui Tang

Scientific
Publishing

Dissertation, Karlsruher Institut für Technologie (KIT)
Fakultät für Bauingenieur-, Geo- und Umweltwissenschaften, 2013
Referenten: PD Dr. Stefan Norra, Univ.-Prof. Dr. Huaming Guo

Impressum

KIT Scientific Publishing

Karlsruher Institut für Technologie (KIT)
KIT Scientific Publishing
Straße am Forum 2
D-76131 Karlsruhe

KIT Scientific Publishing is a registered trademark of Karlsruhe Institute of Technology. Reprint using the book cover is not allowed.

www.ksp.kit.edu

Print on Demand 2014

ISSN 1618-2677
ISBN 978-3-7315-0144-2
DOI: 10.5445/KSP/1000037419

SEPARATINGARSENIC OXYANIONS FROM NATURAL WATERS FOR OXYGEN ISOTOPE ANALYSIS

Zur Erlangung des akademischen Grades eines

DOKTORS DER NATURWISSENSCHAFTEN
(Dr. rer.nat.)

von der Fakultät für

Bauingenieur-, Geo- und Umweltwissenschaften

des Karlsruher Instituts für Technologie (KIT) – Universitätsbereich

genehmigte
DISSERTATION

von
M.Sc. Xiaohui Tang

aus Anqing, China

Tag der mündlichen Prüfung: 06.11.2013
Hauptreferent: PD Dr. Stefan Norra
Korreferent: Univ.-Prof. Dr. Huaming Guo

Karlsruhe 2013

"There is no such thing as a failed experiment, only Experiments with unexpected outcomes."

-Richard Buchminster Fuller (1895-1983)

Kurzfassung

Natürliche Verunreinigung des Grundwassers durch Arsen ist ein entscheidendes Wasserqualitätsproblem in vielen Teilen der Welt. Trotz erheblicher Anstrengungen zum Verständnis seiner Mobilisierung in natürlichen Gewässern, sind die exakten Mechanismen des regionalen Auftretens hoher Arsen-Konzentrationen im Grundwasser noch ungeklärt. Arsen tritt meist in Form von Oxianionen in natürlichen Gewässern auf und die Untersuchung der Sauerstoff-Isotopenzusammensetzung in Arsen-Oxianionen kann neue Einblicke in die Herkunft & Mobilisierung von Arsen im Grundwasser bieten.

In dieser Studie wurden effiziente Methoden entwickelt für die i) quantitative Abtrennung von Arsenspezies von verschiedenen anderen gelösten oxianionischen Spezies, z.B. PO_4^{3-}, HCO_3^-, SO_4^{2-}, bei der die Anwesenheit von NO_3^- jedoch problemstisch nicht kritisch ist, sondern hohe Konzentrationen von Cl^- reduziert werden müssen; ii) quantitative Fällung des abgetrennten AsO_4^{3-}/AsO_3^{3-}; und iii) Messung der $\delta^{18}O$-Werte in den Arsen-Oxianionen; iv) die Kinetik der Sauerstoff-Austausch zwischen Arsen Oxyanionen und Wasser.

Acht Adsorbentien wurden für die Durchführung der Experimente zur Abtrennung von AsO_4^{3-} und AsO_3^{3-} aus natürlichen Wässern verwendet. Während der Trennverfahren werden hohe Hydrogen-Karbonat (HCO_3^-) Konzentrationen durch Einstellen eines pH-Werts unter 4,3 vermischen CO_2. Arsenat und PO_4^{3-} wurden anschließend von AsO_3^{3-}, SO_4^{2-} und Cl^- durch ein Aluminiumsilikat Sorptionsmittel (MAS) quantitativ getrennt. Die anspruchsvolle Aufgabe ist die AsO_4^{3-}-Abtrennung von PO_4^{3-} aufgrund ihres ähnlichen chemischen Verhaltens. Neunundneunzig Prozent des AsO_4^{3-} wurden durch Verwendung des Harzes Amberlite IRA-400 in OH-Form (IRAN) von PO_4^{3-} abgetrennt. Die quantitative AsO_4^{3-} oder AsO_3^{3-}-Abtrennung von SO_4^{2-} und Cl^- wurde durch die beiden basischen

Adsorbentien Yttrium Karbonat (BYC) und Eisenoxid Nano-Partikeln (NFO) erreicht. Arsenat konnte jedoch auch mit porösen Polymerkügelchen, die mit monoklinem, wasserhaltigem Zirkoniumoxid beladenen sind (MHZR), von SO_4^{2-} und Cl^- getrennt werden. Die Ergebnisse der gesamten Trennung von AsO_4^{3-} oder AsO_3^{3-} haben gezeigt, dass NFO aufgrund seiner Adsorptionseigenschaften viel effektiver ist als alle anderen getesteten Adsorbentien. In den gesamten Extraktionsverfahren von AsO_4^{3-} oder AsO_3^{3-} aus natürlichen Wässern, sind die pH-Bedingungen aus der Reaktionslösung der entscheidende Faktor für jeden Schritt der Trennung.

In Anbetracht der geringen Löslichkeit von Ag-Arsenat und Ag-Arsenit, ist die Fällung mit $AgNO_3$ der geeignetste Weg, um Arsen-Oxianionen aus wäßrigen Lösungen quantitativ zu extrahieren. Nach der Fällung mit $AgNO_3$ konnten durch schnelle Anpassung des pH-Wertes auf 7,5 mit Ammoniak, Wiederfindungsraten des extrahierten AsO_4^{3-} oder AsO_3^{3-} als Ag_3AsO_4 und Ag_3AsO_3 von 96,5 ± 5,6 % (n=10) bzw. 94,2 ± 1,3 % (n=5) erreicht werden. Der Ag_3AsO_3 Niederschlag, wurde in einem Vakuumtrockner gefriergetrocknet und in einem Vakuumexsikkator weniger als eine Woche gelagert, da er sonst mit Luft oxidieren würde. Die XRD Ergebnisse zeigen, dass die Fällungprodukte aus Ag_3AsO_4 für die Detektion von Sauerstoff Isotopenzusammensetzung ausreichend rein waren.

Die Messungen der $\delta^{18}O$-Werte in AsO_4^{3-} und AsO_3^{3-} wurden durch Reduktion mit Graphit bei hoher Temperatur und kontinuierlicher Massenspektrometrie (TC-EA-Py-CF-IRMS) durchgeführt. Die Kalibrierung des $\delta^{18}O$-Wert (VSMOW) wurde mit Hilfe der internationalen Standards USGS47 (H_2O mit $\delta^{18}O$-Wert von -19,8 ‰) und VSMOW-2 (H_2O mit $\delta^{18}O$-Wert von 0 ‰) sowie zwei internen Standards LK-2 (Ag_3PO_4 mit $\delta^{18}O$-Wert von 12,1 ‰) und LK-3 (Ag_3PO_4 mit $\delta^{18}O$-Wert von 17,9 ‰) festgelegt. Die Ergebnisse haben gezeigt, dass die KNO_3 Referenzstandards, USGS34 und IAEA N3, für den Durchführung der Kalibrierung nicht geeignet waren, weil der $\delta^{18}O$-Wert in KNO_3 nicht die gleiche Korrelation zwischen experimentellen Werten und Referenzwerten zeigte. Die

Thermolyse von KNO_3 produziert $^{14}N^{16}O(30)$ Gas, das zusammen mit CO(30) detektiert wurde und den endgültigen $\delta^{18}O$ Wert beeinflußte. Der $\delta^{18}O$ Wert von "AsStd" (reines Ag_3AsO_4), welches als interner Standard für die Steuerung der genauen Messung verwandt wurde, war sehr stabil und zeigte nur Abweichungen kleiner als -3,98 ± 0,27 ‰ (n = 19) während der gesamten Messreihe.

Experimente zum kinetischen Austausch von Sauerstoff zwischen AsO_4^{3-} und Wasser haben gezeigt, dass sich das Gleichgewicht bei verschiedenen pH-Bedingungen und Temperaturen in sehr kurzer Zeit einstellt. Das neue Gleichgewicht wurde bei 25 °C und pH 5, sowie bei 70 °C und einem pH-Bereich von 1-10 innerhalb von 3 Stunden erreicht. Die niedrigste Austauschrate wurde bei 25 °C und pH 10 gefunden, wobei die Halbwertszeit für den Isotopenaustausch 38,5 Stunden betrug. Die Austauschrate ist stark abhängig von Temperatur, pH-Wert und Konzentration von Arsen-Oxianionen in der Lösung. Ein viel schnellerer Austausch wurde bei niedrigen pH-Werten beobachtet. Zudem steigt die Austauschrate mit Temperaturerhöhung. Bei systematischer Betrachtung zeigt sich, dass die Reihenfolge der $\delta^{18}O$ Anreicherung unter Arsenatspezies bei einer gegebenen Temperatur $H_3AsO_4 > H_2AsO_4^- > HAsO_4^{2-} > AsO_4^{3-}$ ist. Diese Beobachtung ist ähnlich zu dem, was in einigen anderen Oxianion-Wasser Systemen beobachtet wurde, z.B. in PO_4^{3-}-Wasser Systemen, SO_4^{2-}-Wasser Systemen, und CO_3^{2-}-Wasser Systemen.

Diese Studie hat ein vollständiges Verfahren für die Messung der Sauerstoff-Isotope in Arsen-Oxianionen in natürlichen Wässern entwickelt. Der schnelle Austausch von Sauerstoff-Isotopen zwischen AsO_4^{3-} und Wasser zeigt, dass die Sauerstoff Fraktionierung zwischen AsO_4^{3-} und dem Grundwasser sehr schnell ist und die entwickelten Methoden zur Trennung daher nicht in der Feldarbeit angewandt werden können. Jedoch kann das Trennungsverfahren angewandt werden, um die Sauerstoff-Isotope adsorbierter Arsen-Oxianionen in Sedimenten zu untersuchen, deren Fraktionierungs-Kinetik noch unklar ist. Darüber hinaus können diese Methoden in weiteren Studien zur Trennung und Vor-Konzentrierung von Arsenspezies, von PO_4^{3-}, SO_4^{2-} oder Cl^- aus natürlichen

Wässern für andere geochemische Analysen angewendet werden. Außerdem könnten sie auch angewandt werden, um andere Spurenelement-Oxianionen mit stabilen Sauerstoff-Isotopen aus natürlichen Wässern zu trennen.

Schlüsselwörter: Arsenatspezies, Arsenitspezies, quantitative Trennung, Adsorption, Desorption, Fällung, Silberarsenat (Ag_3AsO_4), Silberarsenit (Ag_3AsO_3), Messung der Sauerstoff-Isotopen, $\delta^{18}O$-Wert, kinetische Austausch

Abstract

Natural contamination of groundwater by arsenic is a crucial water quality problem in many parts of the world. Despite considerable efforts in understanding its mobilization in natural waters, the exact mechanism of the regional occurrence of high arsenic concentration in groundwater is still unclear. Arsenic occurs mostly in the form of oxyanions in natural waters, and the investigation of the isotopic composition of oxygen in As-oxyanions may offer new insights into arsenic source and mobilization in groundwater.

In this study, the efficient methods were developed for i) quantitative separation of arsenic oxyanions from various dissolved oxyanionic species, like PO_4^{3-}, HCO_3^-, SO_4^{2-}, during which the presence of NO_3^- is not critical but the concentration of Cl^-, which must be reduced; ii) quantitative precipitation of AsO_4^{3-}/AsO_3^{3-} from separated effluent; and iii) measurement of the $\delta^{18}O$ values in arsenic oxyanions; iv) the kinetics of oxygen exchange between arsenic oxyanions and water.

Eight adsorbents were used to carry out separation experiments of AsO_4^{3-} and AsO_3^{3-} from natural waters. During the separation procedures, high bicarbonate (HCO_3^-) concentrations are easily removed as CO_2 by adjusting the pH below 4.3. Arsenate and PO_4^{3-} were subsequently quantitatively separated from AsO_3^{3-}, SO_4^{2-} and Cl^- by an alumino-silicate based sorbent (MAS). The challenging task is the separation of AsO_4^{3-} from PO_4^{3-} due to their similar chemical behavior. Ninety-nine percent of AsO_4^{3-} was separated from PO_4^{3-}, by using the resin Amberlite IRA-400 with OH form (IRAN). The quantitative separation of AsO_4^{3-} or AsO_3^{3-} from SO_4^{2-} and Cl^- was achieved by both basic adsorbents basic yttrium carbonate ($Y(OH)CO_3$) (BYC) and nano-particulate iron oxide (NFO). However, AsO_4^{3-} was also separated from SO_4^{2-} and Cl^- by porous polymer beads loaded with monoclinic hydrous zirconium oxide (MHZR). The results of the whole separation of AsO_4^{3-} or AsO_3^{3-} show that NFO is much more effective than others, because of its adsorption properties. In the complete extraction procedures of AsO_4^{3-} or AsO_3^{3-} from natural waters, pH conditions in the reaction solution are the key factor for each step of separation.

Considering the low solubility of Ag-arsenate and Ag-arsenite, the most appropriate way to quantitatively extract arsenic oxyanions from aqueous solutions is to precipitate them with $AgNO_3$. After precipitating with $AgNO_3$ by using ammonia to rapidly adjust the pH to 7.5, the recoveries of the extracted AsO_4^{3-} or AsO_3^{3-} as Ag_3AsO_4 and Ag_3AsO_3 were 96.5±5.6 % (n=10) and 94.2±1.3 % (n=5), respectively. The Ag_3AsO_3 precipitate was freeze-dried in a vacuum drier and stored in a vacuum desiccator for not longer than one week, because it would be easily oxidized by air. The XRD result shows that the precipitates (Ag_3AsO_4) were pure enough for the detection of oxygen isotope composition.

The measurements of the $\delta^{18}O$ value in AsO_4^{3-} and AsO_3^{3-} were performed using high temperature reduction with graphite and continuous flow mass spectrometry (TC-EA-Py-CF-IRMS). The calibration of the $\delta^{18}O$ value (VSMOW) was established using the international standards of USGS47 (H_2O with $\delta^{18}O$ value of -19.8 ‰), VSMOW-2 (H_2O with $\delta^{18}O$ value of 0 ‰), and two internal standards of LK-2 (Ag_3PO_4 with $\delta^{18}O$ value of 12.1 ‰) and LK-3 (Ag_3PO_4 with $\delta^{18}O$ value of 17.9 ‰). The results show that the KNO_3 reference standards, USGS34 and IAEA N3, were not appropriate for establishing the calibration, because the $\delta^{18}O$ value in KNO_3 does not show the same correlation between the experimental values to the reference values. The thermolysis of KNO_3 produces $^{14}N^{16}O$ (30) gas, which was detected together with CO (30) and affected the final $\delta^{18}O$ value. The $\delta^{18}O$ value in the "AsStd", which is pure Ag_3AsO_4 as an internal standard for controlling the precise of measurement, was very stable, showing -3.98 ± 0.27 ‰ (n=19) during whole measurements series.

Experimental results of kinetic exchange of oxygen between AsO_4^{3-} and water show that the entire exchange procedure happens in a very short time at different pH conditions and temperatures. The equilibrium to a new stable state was reached within 3 hours at 25 °C and pH 5, or at 70 °C and pH range of 1-10. The lowest exchange rate was found at 25 °C and pH 10, and the half-time for isotope exchange was 38.5 hours. The exchange rate strongly depends on temperature, pH-value and concentration of arsenic oxyanions in the solution. A much faster exchange was observed at low pH rather than at high pH. It also increases with temperature. From the systematics observed, it seems that the order of ^{18}O enrichment among arsenate species is H_3AsO_4 > $H_2AsO_4^-$ > $HAsO_4^{2-}$ > AsO_4^{3-} at a given temperature. This observation is similar to what has been observed in some other oxyanions-water systems, such as PO_4^{3-}-water systems, SO_4^{2-}-water systems, and CO_3^{2-}-systems.

This study has developed a complete procedure for the measurement of oxygen isotopes in arsenic oxyanions in natural waters. The fast exchange of oxygen isotope between AsO_4^{3-} and water suggests that the oxygen fractionation of AsO_4^{3-} in groundwater is very fast and the developed methods of separation may not be applied to fieldwork. However, the method of separation may be applied to investigate the oxygen isotope of adsorbed arsenic oxyanions in sediments, whose fractionation is still unclear. Furthermore, those methods may be applied as well to further studies on the separation and pre-concentration of arsenic species, PO_4^{3-}, SO_4^{2-} or Cl^- from natural waters for other geochemical analysis. Besides, it might be also applied to separate other trace oxyanions with stable oxygen isotope from natural waters.

Key words: arsenate species, arsenite species, quantitative separation, adsorption, desorption, precipitation, silver arsenate (Ag_3AsO_4), silver arsenite (Ag_3AsO_3), measurement of oxygen isotope, $\delta^{18}O$ value, kinetic exchange

摘要

元素砷存在于世界很多区域的天然地下水中，并导致严重的水质问题从而影响人类的健康。虽然近几十年已经开展了许多对自然界中砷迁移的研究，但是区域高砷地下水的确切形成机理依然没有明确。由于地下水中的大部分砷是以砷酸盐（含氧阴离子）的形式存在，对砷酸盐中的氧同位素开展研究可能会对探讨地下水中砷迁移的机理提供新思路。

在这篇论文中，几种有效的方法被试验出来以用于检测地下水中砷酸盐的氧同位素含量：i）迅速地定量分离出砷酸盐和地下水中普遍存在的其它含氧阴离子，例如 PO_4^{3-}，HCO_3^-，SO_4^{2-}。但是水中存在的 Cl^-离子浓度也被降低。而存在水中的含氧阴离子如 NO_3^-因为其极强的水溶性而不会对之后的实验产生影响。2）成功地从已分离出的砷酸盐溶液中定量沉淀出 AsO_4^{3-}和 AsO_3^{3-}离子。3）开发出有效的测定砷酸盐中的氧同位素的方法。4）明确了砷酸盐和水之间的氧原子交换速率。

8 种不同的吸附剂被用于分离天然水中的 AsO_4^{3-}和 AsO_3^{3-}的试验中。在分离实验中，通过调节水中 pH 值<4.3，高浓度的 HCO_3^-能被迅速反应成为 CO_2 而去除。随即 AsO_4^{3-}和 PO_4^{3-}与水中的 AsO_3^{3-}，SO_4^{2-}，Cl^-被基于铝硅酸盐的吸附剂（MAS）定量分离。而由于 AsO_4^{3-}和 PO_4^{3-}具有极其相似的化学特性，定量分离它们是一个比较具有挑战性的任务。Amberlite IRA-400 树脂（IRAN）能够分离 99%的 AsO_4^{3-}与 PO_4^{3-}。碳酸氢氧钇（$Y(OH)CO_3$）（BYC）和纳米氧化铁（NFO），这两种吸附剂，能够分离水中的 AsO_4^{3-}或者 AsO_3^{3-}与 SO_4^{2-}和 Cl^-。同时，AsO_4^{3-}与 SO_4^{2-}和 Cl^-亦能被载有斜水合氧化锆的多孔聚合物（MHZR）。由于区别于其他两种吸附剂的吸附性能，实验证明 NFO 是最佳分离水中砷酸盐与 SO_4^{2-}和 Cl^-的材料。在整个分离试验中，反应溶液的 pH 值是每个分离步骤中的关键影响因素。

由于砷酸银（Ag_3AsO_4）和亚砷酸银（Ag_3AsO_3）的溶度积非常低，定量提取水溶液中的砷酸盐的最合适方法是利用可溶性银溶液沉淀溶液中的砷酸盐。这个方法的重要反应条件是反应溶液的 PH 值必须处于中性，在 7.0-7.5。迅速用氨水将反应溶液的 pH 值调节到中性范围致使溶液中的 AsO_4^{3-}或 AsO_3^{3-}定量沉淀，它们在烘干后的 Ag_3AsO_4 或者 Ag_3AsO_3 中的回收率分别为 96.5±5.6%（n=10）和 94.2±1.3%（n=5）。由于 Ag_3AsO_3 很容易在空气被氧化，需要在真空干燥器中冷冻干燥，并存储于真空干燥器中不超过一周。 X 射线衍射结果表明，所得沉淀物中的 Ag_3AsO_4 化学成分纯净，并足以用于测试所沉淀的砷酸盐中的氧同位素。

AsO_4^{3-}和 AsO_3^{3-}中的 $\delta^{18}O$ 值（VSMOV）使用高温还原石墨和连续流质谱（TC-EA-Py-CF-IRMS）进行测定。其中使用两种国际标准物质和两种实验室内部标准物质用于建立校准砷酸盐中的 $\delta^{18}O$ 值（VSMOV）的标准曲线。其中国际标准物质是 USGS47（H_2O，$\delta^{18}O$ 值为-19.8‰）和 VSMOW-2（H_2O，$\delta^{18}O$ 值为 0‰），而实验室内部标准物质是 LK-2（Ag_3PO_4，$\delta^{18}O$ 值为 12.1‰）和 LK-3（Ag_3PO_4，$\delta^{18}O$ 值为 17.9‰）。在建立标准曲线的测试结果中显示，由于 KNO_3 中 $\delta^{18}O$ 值的参考值与实验值之间并不显示相同相关性，KNO_3 标准物质（如 USGS34 和 IAEA-N3）并不适用于建立校准曲线。原因是，KNO_3 在热解过程中会产生 $^{14}N^{16}O$（30），这些气体会与 CO（30）气体一起被质谱（MS）检测到从而影响到最终的氧同位素值。纯净的 Ag_3AsO_4 被命名为“AsStd”，其 $\delta^{18}O$ 值在整个测试过程中非常稳定，因此用于作为内部标准进行精确控制的测量样品。它的 $\delta^{18}O$ 平均值在整个测量期间系列中-3.98±0.27‰（n=19）。

氧原子交换动力学实验结果显示，在不同的 pH 值和温度条件下，氧原子在 AsO_4^{3-}与水之间交换速度极快并达到一个新的平衡值。结果显示在 25℃和 pH 值 1 和 5 的条件下，或在 70℃和 pH 值范围 1-10 下，氧原子交换都会在 3 小时内达到一个新的稳定状态的平衡。而相对慢的交换速率发生在 25℃和 pH10 的条件下，其氧同位素交换的半衰时间为 38.5 小时。溶液中的砷酸盐与水的交换速率强烈地依赖于温度，pH 值和浓度。交换速率随 pH 值的升高而减慢，随温度的升高而加快。在整个反应系统的结果中，在给定的温度下，^{18}O 富集在砷酸盐中的顺序为 H_3AsO_4 > $H_2AsO_4^-$ >

$HAsO_4^{2-}$ > AsO_4^{3-}。这个测试结果显示氧同位素在水溶的砷酸盐-水系统的交换行为与一些其他的含氧阴离子-水系统非常相似，例如磷酸盐-水系统，硫酸盐-水系统和碳酸盐-水系统。

在本篇论文中，一个完整的测试天然水中砷酸盐中的氧同位素的方法被试验出来。然而，AsO_4^{3-}和水之间的氧交换实验显示水溶液中的 AsO_4^{3-}中的氧交换速度非常快，由此可以推断地下水 AsO_4^{3-}中的氧分馏也会非常快。因此所试验出的分离方法将不适合应用于野外样品的采集，同时也显示砷酸盐中的氧同位素很难用于追踪地下水中的砷来源。虽然如此，所得出的试验方法也许可以应用于测试吸附在沉积物中的砷酸盐中的氧同位素。而且，也可以应用于天然水中的其他阴离子（例如，PO_4^{3-}，SO_4^{2-}和 Cl^-）的分离富集实验以用于其他地球化学分析。此外，它也可能适用于从天然水域中分离其他微量含氧阴离子以测定他们的氧同位素值。

关键词： 砷酸盐，亚砷酸盐，定量分离，吸附，解吸，沉淀，砷酸银（Ag_3AsO_4），银亚砷酸（Ag_3AsO_3），测量氧同位素 $\delta^{18}O$ 值，动力交换

Contents

Figures

Tables

Abbreviations

- AA0: Analig AS-01
- AsStd： Internal pure Ag_3AsO_4 standard material
- BET: Brunauer–Emmett–Teller
- BYC: Basic Yutrium Carbonate
- DL: detection limit
- DMA: Dimethylarsinic acid
- DW: de-ionized water (18.2 MΩ.cm MilliQ)
- EM: Electronic Microscopy
- FIAS: Fluid injection atomic spectrometry
- h: hour
- HR-ICP-MS: High Resolution-Inductively Coupled Plasma-Mass Spectrometry
- ICP-OES: Inductively Coupled Plasma-Optical Emission Spectrometry
- IRA: Amberlite IRA – 400 (OH) resin
- IRAN: Amberlite IRA – 400 (OH) resin after washing by 0.25 M NaOH
- MAS: an alumino-silicate based sorbent
- MBLO: Muromac B-1-La-OH
- MHZR: the porous polymer beads loaded with monoclinic hydrous zirconium oxide
- Min: minute
- MMA: Monomethylarsonic acid
- NFO: nano-Fe_2O_3 particle
- V-NMDG: Immobilization of N-methyl-D-glucamine (NMDG) on poly - (vinylbenzyl chloride) beads
- pKa: An acid dissociation constant
- rpm: rotation per minute

- SEM-XRF: scanning electron microscopy coupled with an X-ray fluorescence detection device
- TC/EA-GC –IRMS: Thermal Conversion/Elemental Analyzer-Gas Chromatogrphy-isotope ratio mass Spectrometry
- XRD: X-ray Diffraction
- XRF: X-ray fluorescence

Acknowledgements

This doctoral thesis could not have been possible without the help and support of all the kind people around me, to only some of whom it is possible to give particular mention here.

- First of all I want to express my gratitude to Dr. Zsolt Berner, whose project allowed me to start working on my thesis and get it close to finalization. Especially, I want to point out his wide experience in all fields of geosciences and in geochemical laboratory work, from which I could extensively benefit during all these years. I highly appreciate the fruitful discussions, which we had almost every day - not only, but especially - during the difficult moments of my work. But he guided me not only in scientific work, and he was an example to follow also with his life attitude. I wish him a beautiful time and good health in his retirement!

- I am especially grateful to my supervisor PD. Dr. Stefan Norra, without whose support I could not have finalized my thesis. I could always enjoy his fully support in solving my problems regardless whether they concerned my work, my thesis or my stay in Germany. Thanks a lot for his kindness and for his patience with my work.

- I am in deep debt with my reference supervisor Prof. Dr. Huaming Guo, who helped me with proofreading and correcting inherent mistakes and errors in my thesis. He gave me helpful suggestions and supported me also in solving all kind of minor/major problems I met in my work and living during my stay in Germany.

- I want to pay tribute to Prof. Dr. Torsten Vennemann from whose abundant knowledge and long experience in isotopic geochemistry and related laboratory work I could benefit so much during my stay in Lausanne, Switzerland. I am thankful for his bright ideas and useful advices during the discussions we had on interpreting my experimental results.

- I thank Prof. Dr. Thomas Neumann for his kind encouragement and for his support, which allowed me to extend my stay in the Institute of Mineralogy and Geochemistry, KIT.

Further on, I owe a profound a debt of gratitude to:

- Gesine Preuss, for her far-reaching support and patience in introducing me especially in the practice of measurements with TC/EA-IRMS but also with FIAS;

- Dr. Utz Kramar, for his support in carrying out measurements with XRD and XRF;

- Dr. Gerhard Ott, for the grain size measurements with Scanning Electron Microscopy (SEM) and also for some very useful suggestions for my study;
- Dr. Thierry Muller from the Instiute of Organic Chemistry (IOC) and Mr. Nalcaci from the Engler-Bunte-Institute, Division of Combustion Technology (EBI-VBT), for their helps in preparation of different adsorbents;

- Dr. Pirimze Khelashvili, for the many analyses she carried out for me by IC and photospectrometry, and to

- All the technicians at the IMG who gave me a hand in the laboratory work or carried out all kinds of analyses for me: Cornelia Haug (wet chemistry lab), Claudia Mössner (ICP-MS, IC), Beate Oetzel (XRF, XRD), Christian Nikoloski (sample preparation, thin sections);

- My officemates, Dr. Elisabeth Eiche and Dr. Nina Schleicher, who helped me so much and in so many respects since I came to Karlsruhe. They were especially patient in correcting and improving both my English and German. I appreciate the unforgettable time we spent together during my study years at the IMG;

- Mrs. Halina Nytz and Mrs. Andrea Friedrich who were always very nice and friendly in solving different administrative issues and making arrangements for my conference and work trips;

- Dr. Kuo Tian, Andreas Holbach and Alexandra Nothstein for their kindness in correcting the German and English abstracts of my thesis;

- Dr. Harald Neidhardt, Dr. Alexader Diener, Dr. Jan Stelling, my latest office mates Helena and Josi, with whom we had so much fun together even only within the short time.

- All the other friends and PhD fellows from the IMG for the fun together and their help: Wei, Yuan, Peter, Monika, Maren, Dirk, Olga, Arno, Sebastian, Nicolas;

- Dr. Kerstin Bauer and Dr. Kocsis László, the friends I made in Laussane, helping me a lot with my work and living in Lausanne;

- All my Chinese friends in Germany, for the great time we spent together during our stay in Karlsruhe.

Finally, last but not least, my especial thanks are due to my sponsors, who actually made possible to carry out my research:

- The Deutsche Forschungsgemeinschaft (DFG), for funding the research project within which I could work on the topic of my thesis;
- The China Scholarship Council (CSC), who funded my living in Germany for my PhD study;
- Graduiertenschule für Klima und Umwelt (GRACE) and Karlsruhe House of Young Scientists (KHYS), at the KIT, giving me the chance to participate at different international conferences, like in USA, in Budapest, and allowing me a three months stay in Lausanne. Hereto, I want to thank especially Dr. Lucas and Mrs. Engelmann.

Now, I want to thank my family, and all my teachers and friends in China in my mother tongue/ 在此， 我必须要感谢我远在在中国的家人，导师和朋友：

- 我的父母: 唐礼富和钱桂芳，是你们的无私的付出和支持让我能够在德国安心的生活并顺利完成我的论文。是你们的辛勤工作和教育让我能有机会走在世界各地开阔视野。
- 我的舅舅舅妈们：感谢你们一直以来对我的学习生活上给予的经济和精神上的大力支持，在此，我想对你们表达深切的谢意。
- 我的所有亲人：感谢你们对我在德国生活的关心，同时感谢所有兄弟姐妹们一直以来开心欢乐的团聚生活，尤其感谢哥哥嫂嫂妹妹的大力支持。
- 我的硕士导师刘菲老师，感谢您对我学习生活的方向性指导，让我对以后的人生规划有了更多的认识和想法。
- 我的国内的所有好朋友们，虽然不能一一列举你们的名字，但是我会一直感激那些我们一起相处的快乐时光。
- 最后我要感谢我的先生张廓，感谢你在过去两年时间里的快乐相处和无私的支持。没有你的陪伴，我不能如此开心地生活并顺利地完成我的论文。

1 Motivation and objective

1.1 Motivation

After many years of intensive research, there has been a general consensus about the decisive role of reducing conditions in the release of arsenic (As) from sediments into the groundwater in the As contaminated areas of southeast Asia. It is assumed that the reductive dissolution of Fe-(Mn)-oxyhydroxides, with which As is adsorbed to or co-precipitated, is chiefly responsible for the occurrence of the high As concentrations in groundwater. However, the factors, which trigger the changes in redox conditions, are elusive and still under debate. There are several possible factors being considered for the decrease in the redox conditions in sediments, like microbial processes, which could be controlled by the availability of readily degradable organic as a consequence of sewage infiltration, of use of organic fertilizers, of peat layers, of oxbow-lakes or ponds with stagnant disomic waters. Meantime, the intensive groundwater abrasion and particular agricultural practices are currently also invoked to be ultimately the cause of change of the redox conditions in aquifer. However, the expected overlapping of high As contents in groundwater with reducing environments is often missing and the correlations with any of the factors mentioned afore is very often more than equivocal. Unfolding the background for this apparent inconsistency would be of great help in concretizing the factor(s), which ultimately are responsible in the development of reducing conditions in sediments. The knowledge would be influential in implementing efficient strategies to combat this unprecedented calamity. This occasionally observed inconsistency may be explained if it is assumed that As does not always derive locally from the surrounding sediments (i.e., it is autochtonous), but it was transported from a low redox environment to the site of observation (allochtonous).

The investigation of the isotope composition of oxygen in the arsenate/arsenite systems may offer a novel approach to verify this assumption. A comparative study of the isotope composition of the oxygen in water, in As oxyanions dissolved water, and in those adsorbed on the sediment would allow shedding light on the autochthonous or allochthonous characteristics of As in the groundwater. It is expected that analogies with the phosphate system, entitle to assume very low isotope exchange rates between water and arsenite. If this assumption was as what is expected, a similar isotope composition of arsenate/arsenite adsorbed in sediment with that dissolved in groundwater would suggest that As in groundwater is derived directly from the surrounding sediment. In contrast, dissimilarities in isotope composition would imply that desorption of As occurred away from the site where the high As concentrations occur. In a similar way, comparing the δ18O values of arsenate, arsenite and water would allow constraining the timing of redox changes in the groundwater system. The comparison between the two systems would possibly open new perspectives in exploring and understanding the mechanisms that control the geochemical cycling of As. But before that, an appropriate sample preparation and measurement technique for arsenate and arsenite must be worked out.

1.2 Objective of the study

The aim of this thesis is to work out a method to separate, pre-concentrate, extract and measure the isotope composition of the oxygen in As-oxyanions, like arsenate and arsenite. A key issue for the study is the development of a procedure to quantitatively separate and pre-concentrate arsenate and arsenite from low concentration aqueous solutions. Considering the manifold similarities in the chemical behaviours of arsenate and phosphate, it is a hard task to separate arsenate from phosphate in the study. Arsenite, the dominant species of arsenic in groundwater (in many arsenic-affected areas > 80% of arsenic occurs in form of As(III)), might be more important compound to be separated from other ions for the analysis of oxygen isotope. The quantitative precipitation of arsenate and arsenite as solids from solution is another important task for the further measurement of oxygen isotope. One of the major tasks of the study is the accurate measurements of $\delta^{18}O$ value in As-solid phase the establishment of calibration with TC/EA-IRMS.

The kinetics of the isotope exchange between water and arsenite/arsenate is the basic task, which is decisive for the applicability and usefulness of the isotope system of As-oxyanions. Due to manifold similarities between phosphorous and As in respect of their chemical behaviors and chemical bonds, it is reasonable to assume a comparable behavior to phosphate (Lecuyer et al., 1999). It is supposed that the oxygen isotope exchange between arsenite/arsenate and water may be also slow at environmental temperatures that may practically preserve its source-specific isotopic signature for sufficiently long time.

2 State of the art

2.1 Properties and distribution of arsenic

Arsenic (As) is a semi-metal element in the periodic table with atomic number 33, which was first documented by Albertus Magnus in 1250 BC (Emsley, 2011). Arsenic has three common allotropes, metallic gray, yellow and black arsenic, of which gray is the most common and stable form (Norman, 1998). The density of arsenic is 5.727 g/cm^3 and the sublimation point is 615 °C, the triple point is 817 °C (363 MPa) and the critical point is 1400 °C (Gokcen, 1989). As a metalloid listed in group Va of the periodic chart, it exists in nature with four valence states in nature, As(V), As(III), As(0) and As(-III). Arsenic is a monoisotopic element in nature, which is composed of one stable isotope, ^{75}As (Audi et al., 2003).

Arsenic is ubiquitous throughout earth crusts, soil, sediments, water, air and living organisms and comprising about 0.0005 ‰ of the earth's crust (Mandal & Suzuki, 2002; Woolson, 1977). The terrestrial abundance of arsenic is around 1.5-3 mg/kg and the average concentration of arsenic in igneous and sedimentary rocks is 2 mg/kg. In most rocks, arsenic amount ranges from 0.5 to 2.5 mg/kg, and higher concentrations were detected in finer grained argillaceous sediments and phosphorites (Kabata-Pendias, 2000). More than 245 different natural mineral forms containing arsenic, of which approximately 60% are arsenates, 20% sulfides and sulfosalts and the remaining 20% includes arsenides, arsenites, oxides, silicates and elemental arsenic (As). Only certain minerals are commonly encountered in significant amounts (Mandal & Suzuki, 2002; Onishi, 1969). Arsenic in its most recoverable form is in various types of metalliferous deposits and the most common arsenic minerals are arsenopyrite (Mandal & Suzuki, 2002). The weathering of mineral resources makes arsenic mobile on the earth crust during the reaction between the arsenic compound and oxygen in the air.

2.2 The problem of high-arsenic groundwater

As a well-known toxic element in the environment, arsenic causes severe water quality problems in many areas of the world, especially in parts of south and east Asia (Berg et al., 2001; Bhattacharyya et al., 2003; Charlet & Polya, 2006; Smedley & Kinniburgh, 2002; Smith et al., 2000; Stüben et al., 2003), including Argentina, Bangladesh, Bolivia, Brazil, Chile, China, Ghana, Greece, India, Japan, Korea, Mexico, Mongolia, Nepal, New Zealand, Poland, Taiwan, Vietnam, and the USA (Aiuppa et al., 2003; Bibi et al., 2008; Casentini et al., 2011; Kim et al., 2002; Thakur et al., 2011; Yoshizuka et al., 2010). The scale of the problem is notably grave in the Bengal Delta Plain (BDP), including Bangladesh and West-Bengal, India, where the population heavily relies on groundwater with high As concentrations. Whereas, in china, concentrations of As in groundwater of most affected provinces ranged from 220 to 2000 μg/L, while the highest concentration can be 4440 μg/L (He et al., 2012; Mandal & Suzuki, 2002). More than 50 000 habitants are suffered from long term and chronic exposure for As with concentration above 50 μg/L (Jin et al., 2003; Han et al., 2007).

Chronic As toxicity is a global environmental health problem, which has caused serious health problems, including thickening and discoloration of the skin, stomach pain, nausea, vomiting, diarrhea, numbness in hands and feet, partial paralysis, and blindness. It has been also linked to cancers of the bladder, lungs, skin, kidney, nasal passages, liver, and prostate (*eg.* Chung et al., 2013; Goeckerman, 1940; Halim et al., 2009; Hughes et al., 2011; Johnson et al., 2010; Martinez et al., 2011; Ng et al., 2003). In addition to contamination of the drinking water resources, arsenic in groundwater also enters the food chain via irrigation with arsenic-contaminated groundwater (eg, Niethardt, 2012; Norra et al., 2005; Sahoo & Kim, 2013). Arsenic naturally occurs in paddy soil and increases by irrigating with As-rich groundwater, which may have a subsequent impact on rice grains (Sahoo & Kim, 2013). In early 2012, it has been reported that little high concentration of arsenic was detected in juice concentrates and organic brown rice syrup in the production of certain food products because of use of As contaminated water, and the toxic level of As in fruit juice is ~20 μg/L and baby formula is ~300 μg/L ((*Consum Rep.*, 2012; Jackson et al., 2012).

There are different specified rules in different countries for the concentration of arsenic in the drinking water. The former maximum contaminant level (MCL) for all forms of arsenic in groundwater is set to 50 μg/L by EPA in 1975 based on a Public Health Service standard originally established in 1942. However, before the new rule for the concentration of arsenic in drinking water, researchers have concluded that the health risks from arsenic are greater than expected. Studies show that the maximum contaminant level (MCL) for arsenic should be 3-5 μg/L. Because of the cost of achieving this level for all public water supplies would be exorbitant, a 10 μg/L MCL has been adopted in 2001 and subsequently enforced by January 2006 (WHO, 2001; EPA, 2001). There are still some states in the U.S.A, like New Jersey, considering 5 μg/L as the MCL in drinking water. However, in many areas, like Bangladesh, China, and most of Latin American countries, which still suffer arsenic problems in drinking water in some areas, the MCL is defined to be 50 μg/L in rural areas. In Germany and Canada the limit is 25 μg/L (Mondal et al., 2006). In México, the MCL has been 25 μg/L since 2005 (SSA, 2000).

2.3 Mechanisms for the source of arsenic in groundwater

In order to mitigate the problem and to assure a sustainable groundwater supply, the disclosing of the main factors which led to the occurrence of the high As concentrations is urgently needed. Many studies have been conducted to elucidate the cause of the regional occurrence of high arsenic concentrations in groundwater and to work out and implement measures. In order to mitigate its impact on population since the regional scale of the problem became manifest in the early 1990´s (notably in some countries of SE Asia) (BGS & DPHE, 2001; Eiche, 2002; Niethardt, 2012). Despite of the intensive efforts of different scientific, economic and governmental institutions, on national and international scale, the problem is still unsolved and new high As areas have been reported.

The aquatic chemistry of inorganic As(V) and As(III) species is already well known. As(III) is known to be 25-60 times more toxic than As(V) (Silver & Phung, 2005), which always exist as domain As species in groundwater (Guo et al., 2008a). Arsenic(III) and As(V) undergo acid-base equilibria, thus different species are present depending on the pH of water. H_3AsO_3 dissociates sequentially in water to $H_2AsO_3^-$, $HAsO_3^{2-}$, and AsO_3^{3-}, and the undissociated H_3AsO_3 is by far the most dominant species (>99%) at pH close to 7 (Pierce & Moore, 1982; Sharma & Sohn, 2009). In contrast, in similar neutral pH range, As(V) is present as both $H_2AsO_4^-$ and $HAsO_4^{2-}$ in almost equal concentrations. Evidently, the oxidation of As(III) to As(V) is coupled with the uptake of an additional oxygen atom, while the provenance of this atom will substantially influence the bulk isotope composition of the newly formed arsenate complexes.

At present, there is a wide consensus on the natural (non-anthropogenic) source of As in the groundwater. In the source regions of the sediments, arsenic is adsorbed as oxyanion complex onto or co-precipitated with Fe-/Mn-oxides/hydroxides, and transported as particulate matter by rivers to the regions of sedimentation, which becomes an intrinsic part of the aquifer deposits. From the aquifers, arsenic is released into the groundwater mainly by the reductive dissolution of Fe-/Mn-oxides/hydroxides due to a local decrease in the redox potential in the aquifer (Bhattacharyya et al., 2003; BGS & DPHE, 2001; Harvey et al., 2002; Nickson et al., 1998; Smedley & Kinniburgh, 2002; Stüben et al., 2003). This process changes not only the speciation of iron, but also that of arsenic, generally leading to an increase in the relative amount of the more toxic and mobile As(III) relative to As(V) (Smedley & Kinniburgh, 2002; Zobrist et al., 2000). However, the role of human activities in the occurrence of high As concentrations is still a matter of speculations and intensive debate.

Reduction of hydrous iron/manganese oxides is a common process in clastic sedimentary environments. The presence of redox species (like O_2, NO_3^-, SO_4^{2-}, As, Fe and Mn) may put constraints on the thermodynamically possible and microbially favoured processes (e.g., denitrification, Mn(IV)-, Fe(III)- or sulphate-reduction, etc.) which may take place under the prevailing redox conditions. Because many of these redox processes are mediated or accelerated by different groups of microorganisms, the presence of natural organic matter (NOM) and their distribution may control large extent of the As mobility in a sedimentary environment (Charlet & Polya, 2006; Lovley, 1987; 1993; McArthur et al., 2004). McArthur et al. (McArthur et al., 2001) and Ravenscroft et al. (Ravenscroft et al., 2001) were the first to suggest that local changes in redox conditions of the aquifers are possibly related to layers enriched in organic matter (e.g., peat), which would have promoted intensive microbial activity. Some microbial reaction products may also alter the sorption properties of the iron oxyhydroxides (Appelo et al., 2002). Further complicating the redox models can adequately explain the release of As into groundwater.

A competitive exchange of anions (e.g., bicarbonate, phosphate, silicate) in solution with As species (arsenate, arsenite) adsorbed on oxyhydroxides, clay minerals or organic matter may interact with the above mentioned redox processes and lead to an additional desorption of As from mineral surfaces (Holm, 2002; Jain & Loeppert, 2000), though the quantitative role of the competitive exchange is still unclear. Acharyya et al. (1999) has discussed the possible impact of anthropogenic phosphate input, e.g. the excessive use of fertilizers, but no convincing evidences have been reported to support this assumption until now. No correlations were found between high As concentrations in groundwater and agricultural areas treated with P-fertilizers (Bandyopadhyay, 2002; BGS & DPHE, 2001).

The geomicrobiology of As is a very complex issue and the control mechanisms of arsenic cycling during the transformation of organic matter by microbial activity are not yet fully explored and understood. More and more evidences show that the mobilization of As from the solid to aqueous phase is a microbiological phenomenon (Charlet & Polya, 2006; Oremland

& Stolz, 2005) in many aquifers. Micro-organisms may indirectly influence the mobility of As, either by simply altering the redox state of the sediment due to consumption of oxygen during decomposition/mineralization of organic matter (e.g., fermentation, sulphate reduction, etc.) or by using directly Fe(III) as electron acceptor (Fe-(III)-reducing bacteria) and thereby releasing the adsorbed or co-precipitated As (Charlet & Polya, 2006; Lovley, 1993). However, in proximity of oxic/anoxic interfaces, the mobilization of As by enzymatic reduction of Fe(III) may be reversed by aerobic micro-organisms, leading to the re-oxidation of Fe(II) and fixation of As. In a dominant anoxic environment, the presence of metal ions and insoluble sulfides can be formed, which again may serve as sink for As. Experiments with freshly collected sediments from Bangladesh concluded that truly anoxic conditions may not be required for the release of As from reduced, grey colored sediments (Van Geen et al., 2004).

Despite of the considerable progress in understanding the mechanisms, which promote the development of high As groundwaters, some vital questions are still open. The fundamental role of redox changes in the mobilization of As was convincingly documented and is largely accepted, but the factor(s), which finally are responsible for the decrease in the redox potential, still are more in the domain of educated guess than in that of a proved fact. There are some of the invoked redox controls, like infiltration of sewage, of organic rich water from the all-round ponds in these areas, the oscillating redox conditions of rice paddies and other particular agricultural practices, the massive groundwater abrasion, or the occurrence of easily degradable organic matter in aquifer. However, the correlation among these and the redox environment and As content of the groundwater are very often equivocal, without allowing the demonstration of a clear, causal relationship. On the other hand, no efficient protective actions can be worked out and implemented without the firm knowledge of the relative importance of the different circumstances, which lead to the high As concentrations. Furthermore, it is unclear whether the arsenate and arsenite is released right at the site where the high arsenic concentrations occur or rather arsenic is of allochthonous origin and is transported with groundwater to the abstraction site.

This study proposes to develop a completely new approach to tackle with the above-mentioned question by investigating the isotope composition of oxygen in the oxyanion species of As. The identification of the critical factor controlling the redox cycling of As in such areas could make an important contribution to work out effective strategies to manage this calamity of unprecedented scale.

2.4 The technology for arsenic separation

Recently, there are always two purposes for developing separation technologies of arsenic: one is for the detection of concentration of arsenic species in different phases; and another is for the removal of arsenic in contaminated water. The overview here focuses on those methods, which have been already developed for separation of arsenic from other coexisted ions, to disclose some affective materials and methods for preparative and selective extraction of arsenic species from natural water.

2.4.1 The separation of arsenic for concentration detection

An appopriate method for accurately detecting As in water sample is the detection process, which can collect all As information without influences of other coexisted foreign ions. The maximum contaminant level (MCL) of 10 μg/L for arsenic in drinking water (WHO, 2001) requires sensitive and selective detection techniques. Many methods have been developed for detecting total arsenic concentration or different arsenic species concentration in natural water (Ezeh & Harrop, 2013; Melamed, 2005; Narcise et al. 2005; Siegfried et al., 2012; Xiong et a., 2008).

Total As concentration can be analyzed in laboratory by different analytical methods, like Atomic Adsorption Spectroscopy (AAS), Atomic Fluorescence Spectroscopy (AFS), Graphite furnace atomic absorption (GFAA), Hydride generation atomic absorption spectroscopy (HGAAS), inductively coupled plasma-atomic/optical emission spectrometry (ICP-AES/OES), and Inductively coupled plasma-mass spectrometry (ICP-MS). For detection of different arsenic species (As(III), As(V), MMA, DMA) simultaneously in one solution sample, the reaction chromatographic column and the carried liquid phase are very important for separating each specie to flowing out to detector on different times. In recent years, the well-established on-line method for analysis of arsenic species is hyphenation of high performance liquid chromatography (HPLC) with inductively coupled plasma mass spectrometry (ICP-MS) or with Atomic Fluorescence Spectroscopy (AFS), which is based on the use of high performance liquidchromatography (HPLC) and hydride generation (HG) (Garcia-Manyes, 2002; Gómez-Ariza et al., 2000; Liu et al., 2013; Zheng & Hintelmann, 2009).

The relevant methods on analysis of organic and inorganic arsenic compounds have been summarized in Table 2.1 (Liu et al., 2013). It is clear that the delay time of different arsenic specie from the relatively wide compatibility of HPLC mobile phase composition to ICP-MS or AFS detector are dependent on the composition of carried phase and the separation chromatographic column. To detect arsenic species in different types of samples in their specified delay time, there are many compositions of mobile phase are used, like NH_4NO_3, NH_4HCO_3, $(NH_4)H_2PO_4$, H_3PO_4,$(NH_4)NO_3$, HNO_3, organic solutions like methanol/water (1:1), malonate, etc.. The flow rate of the carried phase is between 0.7-1.5 mL/min, and the pH of them must be adjusted in the specified range.

In their analytical system for the selective determination of As(V) and As(III), *Packer and Ciminelli* (2005) used a quaternary amine resin (LC-SAX from Supelco; Bellefonte, PA, USA) as anionic exchanger to pre-concentrate and to separate As(V) from As(III). While As(III) passes unimpeded through the resin, the retained As(V) can be subsequently eluted with 1% (v/v) HNO_3. In a flow-through set-up as described by the authors, separation can be achieved within ca. 3.5 minutes, but the enrichment factor for As(V) was relatively low (ca. 4 times). The higher enrichments of arsenic can be achieved by passing the aqueous sample through a column filled with Muromac A-1, an iminodiacetate-based chelating resin (Muromachi

Chemicals, Japan). This resin can selectively pre-concentrate As(III) after its complexation with pyrrolidine dithiocarbamate (APDC), and the enrichment factors of 136 for As(V) and 160 for As(III) were achieved with 4 min (Karthikeyan & Hirata, 2003). Whereas, Muromac A-1 loaded with La(III) (Trung et al., 2001) can pre-concentrate arsenic in the range of 20 to 30 fold based on the initial concentration when the concentration of As(III), As(V) or the mixture of As(III)+As(V) from 0.5 ppb to 20 ppb and a volume from 50 to 300 ml. This method was used for the preconcentration of trace arsenic species in environmental waters (Trung et al., 2001).

The above mentioned methods are time-consuming and require shipment of water samples to laboratories for speciation determination. Moreover, As(III) can be oxidized to As(V) during transformation and storage of samples from field to laboratory (Borho & Wilderer, 1997). There is one method for detection of arsenic species in the field and laboratories to filter a water sample in a minute with a highly selective alumino-silicate adsorbent, which has already applied for sampling in field (Meng & Wang, 1998; Meng et al., 2001b; Guo et al., 2008b). In this method, only As(III) remains in the filtered samples, and its concentration can be determined by analyzing the total As in the filtered samples. During the filtering process, more than 99 % of As(V) can be removed and 98 % of As(III) is recovered in the filtrates. In the meantime, more han 95% of particulate As in ferric hydroxide suspensions was removed and organic As concentration (i.e. dimethylarsinic acid) can be estimated by filtering oxidized samples and analyzing the total As concentration in the filtrates (Meng & Wang, 1998).

Most of arsenic field tests use mercury bromide, which is extremely toxic, to react with As. For example, the two commercial field test kits, the Arsenator test kit (Wagtech, Palintest, London, U.K) with a test range from 2 µg/L to 100 µg/L and the Merckoquant test kit (Merck, Darmstadt, Germany) with the test range from 0.005 to 3.0 mg/L. Recently, new tests kit to detect As in water sample were developed by using microbiological methods (Siegfried et al., 2012; Ajioka, 2012), which may be an important and main method for the field test because of safeness, low cost, and convenience. One test kit is based on living and Lyophlized bacteria Bioreporters (strain *E.coli* DH5α-2697) emitting bioluminescence as a response to arsenite and arsenite (Siegfried et al., 2012). This test kit was applied during a field campaign in six villages across Bangladesh, and the optimal detection range was defined in range of 0-150 µg/L and the detection limit is 4 µg/L. *Ajioka (2012)* has reported one soil- dwelling bacterium *Baciius subilis*, which can be used as a arsenic biosensor: in the presence of less than 10 µg/L As, they initiate the production of enzymes and an efflux pump for the detoxification and removal of arsenic. The modified bacteria can turn green when arsenic levels are safe and purple when arsenic levels are unsafe.

2.4.2 Removal of arsenic from natural water

Adsorption is the main reaction for removal of arsenic from water. It is a surface phenomenon that was characterized by the concentration of a chemicals species from its vapor phase or from a solution onto or near the surface or pore of solid (Toth, 2002). The interaction between the surface and the adsorbed species may be either chemicals adsorption, in which the forces leading to adsorption are related to chemical bonding forces; or physical adsorption, the adsorption occurs by London-Van Der Waals forces of the solid and adsorbate (Oura et al., 2003). Adsorption is usually described through isotherms, that is, the amount of adsorbate on the adsorbent as a function of its pressure (if gas) or concentration (if liquid) at constant temperature. The quantity adsorbed is nearly always normalized by the mass of the adsorbent to allow comparison of different materials. There are two well-known principle modes of adsorption of molecules on surface: Langmuir (1917) and Freundlich (1906). Langmuir adsorption is an empirical isotherm derived from a proposed kinetic mechanism, which is the simplest theoretical model for monolayer adsorption. Hence this model has been widely used to characterize the adsorption behavior of solutes from aqueous solutions. Freundlich isotherm is an equilibrium isotherm that is used most often in real world examples. The Freundlich equation was developed mainly to allow for an empirical account of the variation in adsorption heat with concentration of the adsorbate (vapor or solute) on an energetically heterogeneous surface.

One type of sorption is ion exchange, which is an exchange of ions between two electrolytes or between an electrolyte solution and a complex (Harland, 1994). In most cases, the term is used to containing solutions with solid polymeric or mineral 'ion exchangers'. Most typical application is preparation of high purity water for power engineering, electronic or nuclear industries.

More than 600 materials were applied for the removal of arsenic in drinking water, groundwater and wastewater (Mohan & Pittman, 2007; Yadanaparthi et al., 2009), including natural materials, biological materials, synthetic adsorbents and commercial ion exchangers. The most removal technologies of arsenic from natural water were oxidation/precipitation, coagulation/coprecipitation, sorption and ion exchange, and membrane technologies (Mohan & Pittman, 2007).

A lot of natural mineral materials can adsorb arsenic in environment, however, the adsorption capacities of them were not so high, like siderite (1.04 mg/g for As(III) and 0.52 mg/g for As(V), (Guo et al., 2007a; 2007b), zeolites (0.017 mg/g for As(III) and 0.1 mg/g for As(V), (Elizalde-González et al., 2001)), Volcanic stone (0.0018 mg/g for As(III), (Elizalde-González et al., 2001)), Geothite (7.5 mg/g for As(III) and 12.5 mg/g for As(V), (Ladeira & Ciminelli, 2004)), Hematite (0.197 mg/g for As(III) and 0.20 mg/g for As(V), (Guo et al., 2007b)), Gibbsite (3.3 mg/g for As(III) and 4.60mg/g for As(V), (Ladeira & Ciminelli, 2004)), Oxisol (2.6 mg/g for As(III) and 3.2 mg/g for As(V), (Ladeira & Ciminelli, 2004)), kaolinite (< 0.23 mg/g for As(V), (Ladeira & Ciminelli, 2004)), etc. Many agricultural products and by-products such as rice husks were also used for arsenic removal from water (Kalid et al., 1998). The industrial by-products/wastes, like Chars and coals (Mohan & Pittman, 2007), red mud (Altundoğan et al., 2000), Blast furnace slag (Ahn et al., 2003), and Fe(III)/Cr(III) hydroxide waste (Namasivayam & Senthilkumar, 1998), were often treated for arsenic removal. Furthermore, some biomaterials, like the fresh and immobilized plant biomass, which can adsorb As(III) without effect of high concentration of Ca and Mg, but with present of Fe(III). The pellets of the white rot fungus *phanerochaete chrysosporium* was used for bioadsorption to remove only As(III) and 100 % Fe(III), and also a little of F^- but very low NO_3^- and Cl^- (Pakshirajan et al., 2013).

Some synthetic adsorbents and ion exchangers for arsenic removal are shown in Table 7.2. Many of them are always with high adsorption capacities of different arsenic species from water. During the adsorption process, the adsorption of arsenic may be induced suppressive effects by precense of other anions, like Cl^-, HCO_3^-, SO_4^{2-}, PO_4^{3-}, Br^-, F^-, etc. The most adsorption mechanism for arsenic follows the inner-sphere complex mechanism.

Many synthetic adsorbents were oxides, like ZrO_2 (Cui et al., 2012), MnO_2 (Lenoble et al., 2004; Zhang & Sun, 2013), hydrous Fe_2O_3 (Jang et al., 2007), CuO (Reddy et al., 2013), have effective adsorption of arsenic. The adsorption capacities of the porous polymer beads loaded with monoclinic hydrous zirconium oxide, for As(V) and As(III) were 89.90 mg/g and 112.4 mg/g, respectively (Suzuki et al., 2000; 2001). The optimal pH range for adsorbing arsnic were 4-6 for As(V) and 9-10 for As(III) without affection of coexisted SO_4^{2-} or Cl^-. *Yuchi et al.* (2003) reported one type of chelating polymer gel loaded with Zr(IV), which can adsorb both As(V) and P(V), and afterwards both of them were stripped from the gel in a KOH solution. However, conditions for a selective stripping of arsenate and phosphate are not given (Yuchi et al., 2003). The amorphous zirconium oxide nanoparticles (Cui et al., 2012) was proven for effective removal of arsenic from aqueous environment without affecteion of the present of Cl^- and SO_4^{2-}, but strongly competing by PO_4^{3-} anion. Manganese oxide (MnO)-loaded polystyrene resin (Lenoble et al., 2004) can adsorb both As(III) and As(V) in 2 hours, but no report about the fate of other coexisted anions. Nanoscale zero-valent iron (Kanel et al., 2005; Rahmani et al., 2011) can affectively remove As(III) from groundwater in the pH range of 4-9, and without interference of NO_3^-, SO_4^{2-}, but strongly competing by other anions, like HCO_3^-, $H_4SiO_4^0$, and $H_2PO_4^{2-}$. Copper oxide (CuO) nanoparticle (Reddy et al., 2013) adsorbed arsenic rapidly and complete adsorptionoccurred only in

30 min. It has the higher capacity at the lower concentration of arsenic; moreover, arsenic adsorption in per unit of nanoparticle was decreased with increasing concentrations of nanoparticle in solution (Reddy et al., 2013). Therefore, the removal of arsenic by CuO nanoparticles is more favorable at a lower concentration than a higher one.

The complex of N-methyl-D-glucamine on poly (Dambies et al., 2004) was proved to selectively adsorb As(V) from the solution with the present of high concentration of PO_4^{3-}, SO_4^{2-} or Cl^-. Fe-Mn binary oxide (Zhang et al., 2007a; 2007b), which has high uptake capability for both As(V) and As(III). During the adsorption on Fe-Mn binary oxide, PO_4^{3-} was the greatest competitor with arsenic for adsorptive sites, but the presence of SO_4^{2-} and humic acid was reported no significant effect on arsenic removal.

Some commercial ion exchangers were also applied for removal of arsenic. Amberlite IRA-400 (Cl) was used for As(V) removal from water, and the adsorption capacity was 27.1 mg/g (Rivas & Muñoz, 2009). The resins Dow XFS-4195A (Chanda, et al., 1988a)and Chelex 100 (Chanda, et al., 1988b) have been proved to selectively remove AsO_4^{3-} and AsO_3^{3-} from Cl^-, SO_4^{2-} by ligand exchange sorption on chelating resins in ferric ion form. The sorption of As(V) anions on Dow XFS-4195A and Chelex was maximum at pH~2 and ~5, respectively, while that of As(III) anions at pH~10 for both resins. The adsorbed anionic species can be readily stripped with dilute NaOH into a concentrated form for recovery. In column operation an enrichment of 40 (Chelex 100) and 100-fold (XFS-4195) relative to influent concentration was obtained for arsenate, as compared to less than 20-fold enrichment for arsenite. The significant phenomenon was both of Cl^- and SO_4^{2-} in high concentration can increase largely the sorption of As(V) on chelex 100(Chanda et al., 1988b) but largely decrease the sorption of As(V) and As(III) on XFS (Chanda et al., 1988a). A combination of solid phase extraction (SPE) (Rahman et al., 2011) columns was tested for quantitative separation of As(III), As(V), MMA and DMA at flow rate of 0.2 mL/min, which was achieved based on the differences in extraction and recovery behavior of the (molecular recognition technology) MRT gel SPE columns with pH for different arsenic species. Likely interference from the various coexisting ions (Na^+, K^+, Ca^{2+}, Mg^{2+}, Cl^-, NO_3^-, CH_3COO^-, PO_4^{3-}, SO_4^{2-}, ClO_4^-) (10 mM) was negligible (IBC Advanced Technologies, 2006; 2007; 2009).

During the removal process of arsenic in artificial or natural water, the fate of the coexisted anions was not reported. Despite the effects of some coexisting anions, like Cl^-, SO_4^{2-}, PO_4^{3-}, SiO_3^{2-}, on the adsorption of arsenic, have been studied, the change of the concentration of those anions in water is not clear in this study, which is very important for the further measurement of oxygen isotope in arsenic oxyanions.

Table 2.1: Summary of relevant methods on analysis of organic and inorganic arsenic compounds in the literature (Liu et al., 2013)

Sample matrix	Extraction	Analytes	Chromatographic column	Mobile phase				*References*
				Composition	Flow rate (mL/min)	pH	Injection volume (μL)	
Animal feed	Methanol/water (1:1)	As(III), As(V), DMA, MMA, ASA, ROX, 4-hydroxyphenylarsonic acid	Agilent ZORBAX SB-C18 (150×4.6 mm, 3.5 μm)	Methanol/water (1:1) mixture and 0.1% formic acid, with tetrabu-tylammonium hydroxide as ion-pair reagent	0.7–1.0	6.25	-	*(Wang et al., 2010)*
Plant material	Methanol/water (9:1)	As(III), As(V), DMA, MMA	Hamilton PRP-X100 (250*4.6 mm, 10 μm)	20 mM $(NH_4)H_2PO_4$ solution	1.5	5.6	25	*(Zheng & Hintelmann, 2009)*
Water	Solid-phase extraction cartridge using anion exchange resin	As(III), As(V), DMA, MMA, ROX	Supelco LC-SAX (4×50 mm); Hamilton PRP-X100 (250×4.1 mm); Dionex AS7 (4×250 mm)	12.5 mM malonate and 17.5 mM acetate; 38–75 mM phosphate buffer; 2.5–50 mM nitric acid in 0.5% methanol	1.0	4.8 5.7	100	*(Bednar et al., 2004)*
White wine	-	As(III), As(V), DMA, MMA	Hamilton PRP-X100 (250×4.1 mm)	20 mM $(NH_4)H_2PO_4$ solution	1.5	6.0	200	*(Moreira et al., 2011)*
Seaweed	Methanol/water (1:1)	As(III), As(V), DMA, MMA, arsenobetaine,arsenosugars	Hamilton PRP-X100 (250*4.1 mm, 10 μm)	20 mM $(NH_4)H_2PO_4$; 20 mM NH_4HCO_3; 20 mM pyridine	1.0	7.0 7.5 2.4	50	*(Van Hulle et al., 2002)*

Sample matrix	Extraction	Analytes	Chromatographic column	Mobile phase				*References*
				Composition	Flow rate (mL/min)	pH	Injection volume (µL)	
Rice, fish meat, chicken meat, and soil	Methanol/water (1:1); 1 M H_3PO_4	As(III), As(V), DMA, MMA, arsenobetaine	Hamilton PRP-X100 (250×4.1 mm, 10 µm); Hamilton PRP-X200 (250×4.1 mm, 10 µm)	10 mM $NH_4H_2PO_4$; 4 mM pyridine/formiate	1.5	6.0 2.8	100	*(Pizarro et al. 2003)*
Marine biological tissues	Methanol/water; 2% HNO_3	As(III), As(V), DMA, MMA, arsenobetaine, arsenocholine, trimethylarsine oxide, tetramethylarsonium	Hamilton PRP-X200; Hamilton PRP-X100	20 mM pyridine; 0.02 mM H_3PO_4; 4 and 60 mM NH_4NO_3	1.5; 1.0	2.7 6.0 8.65	-	*(Whaley-Martin et al., 2012)*
Urine	–	As(III), As(V), DMA, MMA, arsenobetaine, arsenocholine, dimethylthioarsinate, trimethylarsine oxide	ICSep ION-120 (120×4.6 mm, 10 µm)	0 and 100 mM $(NH_4)_2CO_3$ in aqueous 3% (v/v) methanol	0.7; 1.0; 1.5	10.3	25	*(Cubadda et al., 2012)*
Zooplankton and phytoplankton	Water	As(III), As(V), DMA, MMA, arsenobetaine, arsenocholine, trimethylarsine oxide	Hamilton PRP-X100 (150×4.6 mm, 10 µm); Varian IonoSphere C (150 or 100×3 mm, 5 µm)	4 and 60 mM NH_4NO_3; 20 mM pyridine	1.0; 1.5	8.7 2.7		*(Caumette et al., 2011)*
Earthworm	Deionized water	As(III), As(V), DMA, MMA, arsenobetaine, arsenosugar, trimethylarsine oxide	Hamilton PRP-X100 (150×4.1 and 4.6 mm, 10 µm); Varian IonoSphere C (100×3 mm, 5 µm)	20 mM $NH_4H_2PO_4$; 4 and 60 mM NH_4NO_3; 20 mM pyridine	1.0	6.0 8.65 2.7	100	*(Button et al., 2011)*
Seafood	Methanol/water	As(III), As(V), DMA, MMA, arsenobetaine, trimethylarsine oxide	Dionex IonPac AS7 (250×4 mm)	0.5 and 50 mM HNO_3; 0.05 mM benzene-1, 2-disulfonic acid dipotassium salt	1.0	3.4	50	*(Nam et al., 2010)*

Table 2.2: Summary of relevant materials for *adsorption* of different arsenic compounds.

Adsorbent	Type of water	pH range for reaction		Temperature (°C)	Capacity (mg/g) (Langmuir Model)		Reaction time		Affection from the coexisted anions		Regenerator	*References*
		As(III)	As(V)		As(III)	As(V)	As(III)	As(V)	No interference	Competition adsorption		
Alumina, iron hydroxide coated	Drinking water	6.62-6.74	7.15-7.2	30	7.64	36.64	-	-	-	-	-	*(Hlavay & Polyák, 2005)*
Cellulose (bead) with iron oxyhydroxide	Ground water	7.0	7.0	25±0.5	33.2	33.2	-	-	SO_4^{2-}	PO_4^{3-}, SiO_3^{2-}	2 M NaOH	*(Guo & Chen, 2005)*
Fe-Mn binary oxide	Laboratory test	4.8	4.8	25±1	132.6	73.4	24 hours	24 hours	SO_4^{2-}, humic acid	PO_4^{3-}, SiO_3^{2-}	-	*(Zhang et al., 2007a; 2007b)*
FePO4 (amorphous)	Laboratory test	7.0-9.0	6-6.7	20	21	10	-	-	-	-	-	*(Lenoble et al., 2005)*
FePO4 (cryst.)	Laboratory test	7.0-9.0	6-6.7	20	16	9	-	-	-	-	-	*(Lenoble et al., 2005)*
Iron(III)-loaded chelating resin	Aqueous solution	9.0	3.5	25	62.9	55.44	-	-	-	-	0.1 M NaOH	*(Matsunaga et al., 1996)*
Sponge, Fe loaded	Laboratory test	9.1	4.5	75	0.24±0.02 mmol/g	1.83±0.04 mmol/g	24 hours	24 hours	-	PO_4^{3-}>SO_4^{2-}>NO_3^->Cl^-	-	*(Muñoz et al., 2002)*
Nnao Zero-vaent iron	Ground water	7	-	25	3.5	-	1 hour	-	SO_4^{2-}, NO_3^-	HCO_3^-, $H_4SiO_4^0$, $H_2PO_4^{2-}$	-	*(Kanel et al., 2005; Rahmani et al., 2011)*
MnO -loaded polystyrene resin	Laboratory test	3.9-4.3	3.9-4.3	-	52.4	21.9	2 hours	2 hours	-	-	0.2 M NaOH	*(Lenoble et al., 2004)*

Adsorbent	Type of water	pH range for reaction		Temperature (°C)	Capacity (mg/g) (Langmuir Model)		Reaction time		Affection from the coexisted anions		Regenerator	References
		As(III)	As(V)		As(III)	As(V)	As(III)	As(V)	No interference	Competition adsorption		
CuO	Laboratory test	6.39-8.03	6.39-8.03	-	0.42 mg/g in low initial concentration and 0.024 in high initial concentration		30 min	30 min	-	-	0.3 M NaOH	*(Reddy et al., 2013)*
Zirconium oxide, monoclinic hydrous	Drinking water	9.0-10.0	4.0-6.0	25	112.4	89.9	-	3 hours	SO_4^{2-}, Cl^-, NO_3^-, CH_3COO^-	PO_4^{3-}, F^-	1 M NaOH	*(Suzuki et al., 2001)*
Zr resin	Drinking water	8.0	4.5	25	79.42	53.94	-	-	SO_4^{2-}, Cl^-, NO_3^-, CH_3COO^-	PO_4^{3-}, F^-	1 M NaOH	*(Suzuki et al., 2000)*
Zr(IV)-loaded chelating resin	Spring water	9.0	4.0	25	49.15	88.73	-	-	SO_4^{2-}, Cl^-, SO_3^{2-}, NO_3^-, NO_2^-	PO_4^{3-}	1 M NaOH	*(Balaji et al., 2005)*
Amorphous ZrO_2 nanosizes	Aqueous solution	7±0.3	7±0.3	25	>83	>32.4	30 min	30 min	SO_4^{2-}, Cl^-	PO_4^{3-}	1 M NaOH	*(Cui et al., 2012)*
Basic yttrium carbonate	Laboratory test	9.8-10.5	7.5-9.0	40	5.71 mmol/g	6.452 mmol/g	15 hours	1 hour	SO_4^{2-}, Cl^-, Br^- NO_3^-,	PO_4^{3-}	1 M NaOH, 0.5 M acid and Urea	*(Haran et al., 1997; Wasay et al., 1996)*
Chelex 100	Laboratory test	2	10	24	70	45	20 hours	20 hours	-	Presence of Cl^-, SO_4^{2-} increase the adsorption of As(V) and As(III)	$FeCl_3$ solution (pH~2)	*(Chanda et al., 1988a)*
XFS-4195	Laboratory test	8-11 (Max. ~5)	4-7 (Max. ~5)	24	51	46	20 hours	20 hours		Cl^-, SO_4^{2-}	$FeCl_3$ solution (pH~2)	*(Chanda et al., 1988b)*
IRA-400 (Cl^-)	Laboratory test	-	8	20	-	27.1	1 hour	1 hour	-	-	1 M HNO3	*(Rivas & Muñoz, 2009)*
NMDG	Laboratory test	-	6.5	20	-	63.1			Cl^-, SO_4^{2-}, PO_4^{3-}	-	-	*(Dambies et al., 2004)*

2.5 Oxygen isotope system

Oxygen is the most abundant chemical element on the Earth, occurring as elemental O_2, as one of the elements of water, and as functional principle in gaseous, liquid and solid compounds, most of which are thermally stable over large temperature ranges. There are three stable isotopes of oxygen in natural as ^{16}O, ^{17}O and ^{18}O, with which the mean relative natural abundances in [atome-%] are 99.757%, 0.038% 0.205%, respectively (Rosman & Taylor, 1998). Because of the higher abundance and the greater mass difference, the $^{18}O/^{16}O$ ratio is normally determined, which may vary in natural samples by about 10% or in absolute numbers from about 1:475 to 1:525 (Hoefs, 2009).

The $\delta^{18}O$ value of the bulk Earth is constrained by the composition of lunar basalt and bulk chondritic meteorites to be closed to 6 ‰. The dissolved O_2 in the oceans is enriched in ^{18}O relative to atmospheric oxygen, which was demonstrated by *Rakestraw et al.* (1951). The extreme enrichments up to 14 ‰ occur in the oxygen minimum region of the deep ocean due to preferential consumption of ^{16}O by bacteria abyssal ocean waters (Kroopnick & Craig, 1972). The general agreement is that continental glaciation and deglaciation induce changes in the $\delta^{18}O$ value of the ocean on short time scales; however, there is also considerable debate about long-term changes (Hoefs, 2009). The $\delta^{18}O$ value in atmospheric oxygen is rather constant, with 23.5 %, which is produced by photosynthesis without fractionation with respect to the substrate water (Hoefs, 2009).

There are two different δ-scale, $\delta^{18}O$ (PDB) and $\delta^{18}O$ (VSMOW) because of two different categories of users, who have traditionally been engaged in O-isotope studies. The VPDB (Pee Dee Formation) standard is used in low-temperature studies of carbonate, the original supply of which has long been exhausted. Therefore, the secondary standards VSMOW (Vienna Standard Mean Ocean Water) was introduced, whose isotope compositions were calibrated relative to PDB (Hoefs, 2009). The conversion equations of $\delta^{18}O$ (PDB) versus $\delta^{18}O$ (VSMOW) and vice versa (Coplen et al., 1983) are:

$$\delta^{18}O\ (VSMOW) = 1.03091\ \delta^{18}O\ (PDB) + 30.91$$

And

$$\delta^{18}O\ (PDB) = 0.97002\ \delta^{18}O\ (VSMOW) - 29.98.$$

2.5.1 Evaluation of $\delta^{18}O$ value

There are several methods for determination of $\delta^{18}O$ value in some substances, which are the important tools in geological disciplines. The mostly conventional or classical techniques include the thermal decomposition of organic (sometimes also inorganic) compounds in the presence of $HgCl_2$ or $Hg(CN)_2$ or pyrolysis in Ni- or quartz-tubes, as well as the fluorination of silicates. Sulfate and phosphate samples are usually reacted with graphite. Aqueous samples are equilibrated with CO_2 or can be reacted with guanidine hydrochloride, and carbonate samples are digested with concentrated phosphoric acid (Werner, 2003).

The more general applications are achieved by using the online measurement of $^{18}O/^{16}O$ ratios, which is coupling elemental analyzers or gas chromatographs to isotope ratio mass spectrometers. Here, the underlying principle is a reductive reaction with carbon ("carbon reduction") (Belcher et al., 1969), and a prerequisite for this online technique is the fast and quantitative conversion of all 'oxygen' to a single oxygen-containing gaseous products 'CO' (Santrock & Hayes, 1987; Taylor & Chen, 1970).

The Thermo-Chemical Elemental Analyzer (TC/EA) is a pyrolysis (pyro = heat, lysis = to break apart) device that thermally decomposes a sample in a carbon reducing environment in order to break down waters, organic, and inorganic samples (nitrates, phosphates, and sulfates). The early EA-IRMS set-up a reaction zone temperature of 1080 °C was used with glassy carbon grit filling in a quartz tube as the reaction zone (Koziet, 1997). A characteristic of this method is that at the applied temperature glassy carbon cannot convert the oxygen of water to CO, which is an advantage in case of measuring $\delta^{18}O$ value of hygroscopic carbohydrates and other related compounds (Kornexl et al., 1999b). *Kornexl et al.* (1999a) reported that the reaction zone temperature of EA-IRMS must be at 1450 °C, which is very important for the quantitative conversion of oxygen in organic matter and selected inorganic substances like nitrates, phosphates, and sulfates and other oxides into CO. However, for quantitative decomposition of carbonate-O to CO special additives were needed, and the $\delta^{18}O$ of silicates cannot be measured this way as the CO yield was less than 20 %. Higher temperature is needed for quantitatively releasing oxygen from carbonate or silicate minerals by means of the carbon reduction method, respectively. The reaction temperature must be higher than 1600 °C (below that temperature CO is an oxidant to CaC_2 (Hollemann &Wiberg, 1985)) for carbonate minerals and even higher than 2000 °C is required for silicates minerals (Schwander, 1953). Therefore, The classical fluorination of silicates and related substance originally is still very promising and commonly used (Clayton & Mayeda, 1963), even recently the application of laser ablation fluorination systems is the common method for oxygen isotope analyses of silicates minerals (eg. Bindeman et al., 2013; Clayton, 1961; Lester et al., 2013).

2.5.2 Determination of $\delta^{18}O$ value in phosphate from natural samples

The measurement of $\delta^{18}O$ value in phosphate has been well established (Kornexl et al, 1999b; LaPorte et al., 2009; Lecuyer et al., 2007; McLaughlin et al., 2004; Vennemann et al., 2002), by using TC/EA-IRMS. With this method, phosphates are first dissolved as a homogenous solution, which is a very important procedure during the preparation of sample for oxygen isotope analysis. Then they are precipitated as silver phosphate, which soon became the analytic of choice because it is easier to prepare than $BiPO_4$ and, very importantly, because it is not hygroscopic (O'Neil et al., 1994). Silver phosphate is then pyrlyzed reduces with C as CO followed by GC separation of the gaseous products, and their introduction into multiple-collector in the mass spectrometer (MS) for detection of mass of CO with different type of O isotope (Kornexl et al., 1999b).

Due to its relative robustness to diagenetic alteration, the isotope composition of oxygen in phosphate became beside carbonate a valuable complementary tool in assessing climatic conditions in the geological past. Under inorganic conditions kinetics of isotope exchange with water oxygen are 10 times slower in phosphate than in carbonate (Zazzo et al., 2004). On the opposite, the exchange of oxygen isotopes is at least 2 to 15 times faster in phosphate than in carbonate during biologically mediated reactions. During formation of biogenic apatite, oxygen bonds in phosphate are broken through enzymatic reactions and complete isotopic exchange with ambient body water occurs (Blake et al., 1997; Kolodny et al., 1983). In contrast to carbonate-oxygen system, which is currently used to determine the temperature of ambient water with ecotherm organisms, oxygen isotope value in phosphate are increasingly used to assess the isotope composition of ancient meteoric water, which ultimately may provide information about the terrestrial palaeoclimate. Such applications are based on the fundamental observation (Longinelli, 1984) that biogenic phosphate (e.g., in bones or teeth of mammal) is precipitated in new oxygen isotope equilibrium with ambient aqueous (body) fluids. The $^{18}O/^{16}O$ ratio of biogenic phosphate is ultimately controlled by the mean isotopic composition of the meteoric water taken up with nutrition (Longinelli, 1984).

Under abiotic conditions and earth surface temperature (20-45°C) and pressure, both phosphate minerals and dissolved phosphate are resistant to oxygen isotope exchange with water (Blake et al., 1997; Lecuyer et al., 1999; Liang & Blake, 2007). At higher temperatures (70-80°C), exchange kinetics depends on pH, which controls the degree of protonation of the aqueous phosphate species, and lower pH values corresponding to higher exchange rates (O'Neil et al., 2003).

It is well known that enzymes are an essential requirement for catalysis of oxygen isotope exchange between phosphate and water at temperatures < 80°C (Blake et al., 1997; Kolodny et al., 1983). However, it was only recently pointed out that the $\delta^{18}O$ value of dissolved phosphate can be used as a reliable indicator of enzymatic activity and microbial metabolism of phosphate in natural aquatic systems (Blake, 2005; Blake et al., 2001; Colman et al., 2002; 2005). Microbial mediated oxygen isotope exchange between phosphate and water is reflected in a strong positive correlation between the $\delta^{18}O$ values of water and of dissolved phosphate. The microbial metabolism of P was demonstrated by equilibrium oxygen isotope fractionations, which varied with temperature in a manner similar (Blake et al., 2001). it was observed for biogenic apatites that the isotope exchange is independent of the nature (organic or inorganic) or initial $\delta^{18}O$ value of the source of P (Blake et al., 2001). The oxygen isotope composition of the metabolized phosphate records the temperature of equilibration, in conjunction with measured or inferred temperature and the isotope composition of water-oxygen. The $\delta^{18}O$ values of dissolved and sediment associated phosphate can be used to detect microbial enzymatic activity. This approach, which was already successfully used in both marine and terrestrial aquatic environments (Blake et al., 2001; Colman et al., 2002; Liang & Blake, 2005), can detect the role of microbial activity in the transfer of As from sediment to groundwater.

Lately, attempts were undertaken also to use the isotope composition of oxygen in phosphate to trace and quantify the contribution of different sources of phosphate in agro-ecosystems. Mineral fertilizers of marine origin are enriched in ^{18}O relative to dissolved phosphate of "natural" origin (Stern & Wang, 2002). Preliminary results of (Tamburini et al., 2010) indicate that soils amended with mineral fertilizers and composted manure present significantly distinct isotopic signatures, differing in their $\delta^{18}O$ values as much as 6 ‰.

A thorough literature research has shown that, in contrast to phosphate, almost no published information exists on the isotope composition of oxygen in the oxyanionic compounds of As. On the other hand, due to the manifold chemical similarities between P and As, it is reasonable to assume, that the isotope composition of oxygen in these two systems must present some analogies.

3 Methodic design

3.1 Research scheme

The study of the $\delta^{18}O$ value in AsO_4^{3-} and AsO_3^{3-} may be potentially highly instructive in understanding the mechanism of arsenic mobilization in groundwater. The arsenic oxyanions must be separated and quantitatively extracted from natural water for measurement of oxygen isotopic composition. The aim of this study is the development of an efficient method for separation, pre-concentration and precipitation of arsenic oxyanions from other common oxygen bearing anions occurring in natural water, like PO_4^{3-} , HCO_3^-, and SO_4^{2-}. Arsenate and AsO_3^{3-} can be most conveniently precipitated from solution in form of insoluble silver salts (Vogel, 1961), as Ag_3AsO_4 and Ag_3AsO_3, by using $AgNO_3$ solution, of which the presence of NO_3^- anions is not critical. On the other hand, the procedure must allow reducing the concentration of Cl^- ions in order to avoid the formation of large amounts of AgCl in the precipitate. Furthermore, the measurement of $\delta^{18}O$ value in the arsenic oxyanions must be also established in this study.

The main approach of preparative separation of arsenic oxyanions (AsO_4^{3-} and AsO_3^{3-}) for oxygen isotope work is structured as Figure 3.1. There are three important steps, which must be clarified, including separation of arsenic oxyanions from nature groundwater, precipitation of arsenic oxyanions as solid phase, and measurement of $\delta^{18}O$ value in arsenic solid phase. The methods for each step must be first built indoor in order to apply to sampling in fieldwork.

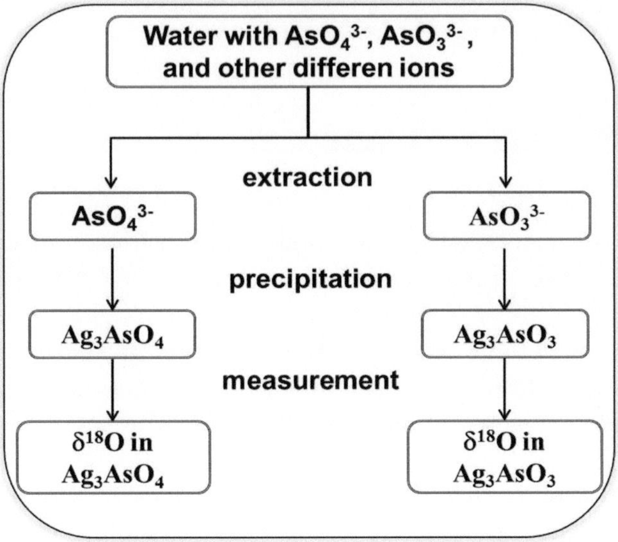

Figure 3.1: The experimental scheme for methodic development of oxygen isotope composition in arsenic oxyanions.

3.2 Thermodynamic calculation

3.2.1 Adsorption isotherm

Adsorption isotherm is a mathematical equation, which describes the relationship between the amount of adsorbate adsorbed on the adsorbent and concentration of the dissolved adsorbate in the liquid at equilibrium at any fixed temperature. The amount of the adsorbed anions q_{eq} (mg/g) at equilibrium is given by:

$$q_{eq} = (C_0 - C_{eq})\, v/m \qquad (1)$$

Where: C_0 (mg/L) is the initial ion concentration, C_{eq} (mg/L) is the concentration at equilibrium, v (L) is the volume of the solution and m the mass of the adsorbent (g) added.

Two isotherm models, Langmuir (1916) and the Freundlich (1906), were used for characterizing the data obtained from the adsorption isotherm, to estimate the adsorption constants, adsorption densities and adsorption maxima for the tested materials.

Langmuir isotherm (1916), one of basically two well-established types of adsorption isotherms, relates the coverage or adsorption of molecules on a solid surface to gas pressure or concentration of a medium above the solid surface at a fixed temperature. The adsorption equation in solution phase is stated as:

$$q_{eq} = q_m \frac{k_l c_{eq}}{1+k_l c_{eq}} \qquad (1)$$

which can be written in a linearized form as

$$\frac{c_{eq}}{q_{eq}} = \frac{1}{q_m} c_{eq} + \frac{1}{q_m} \frac{1}{k_l} \qquad (2)$$

Where: C_{eq} (mg/L) is the concentration at equilibrium, q_{eq} (mg/g) is the amount adsorbed at equilibrium, k_l is Langmuir equilibrium constant, and q_m (mg/g) is maximum amount adsorbed as c_{eq} increases.

Freundlich isotherm (1906), which is another basically well-established adsorption isotherm, relates the concentration of a solute on the surface of an adsorbent, to the concentration of the solute in the liquid with which it is in contact. It applies to both multilayer sorption and non-ideal sorption on heterogeneous surfaces. The Freundlich isotherm is mathematically expressed by the equation:

$$q_{eq} = k_f c_{eq}^{1/n} \qquad (3)$$

which can be written in a linearized form as:

$$\log q_{eq} = log k_f + {}^{1}/_{n}\, log c_{eq} \qquad (4)$$

where n and k_f are the Freundlich constants. n and k_f are the Freundlich constants, whereby k_f is a measure of the adsorption capacity and n a measure of the adsorption intensity (Tanada et al., 1990). The Freundlich isotherm applies to both multilayer sorption and non-ideal sorption on heterogeneous surfaces. The close fit of the experimental data with the Freundlich isotherm suggests the heterogeneity of the sorbent surface. The magnitude of the exponent $1/n$ gives an indication of the adequacy and favorability of the adsorbent/adsorbate system. In most cases, exponent values between $1 < n < 10$ are indicative for a beneficial adsorption (Treybal, 1988). When $1/n$ values are in the range of 0.1 to 0.5, it is easy for adsorbate to be adsorbed (Hassler, 1951). The significant adsorption takes place at low concentration, but as concentration increase the amount adsorbed becomes less significant (Akpomie et al., 2012; Teng & Hsieh, 1998).

3.2.2 PHREEQC Program

PHREEQC is a computer program for simulating chemical reactions and transport processes in natural or polluted water to perform a wide variety of low-temperature aqueous geochemical calculations (Parkhurst & Appelo, 1999). The program is based on an ion - association aqueous model and has capabilities for (1) speciation and saturation-index calculations; (2) batch-reaction and one-dimensional (1D) transport calculations involving reversible reactions, which include aqueous, mineral, gas, solid-solution, surface-complexation, and ion-exchange equilibria, and irreversible reactions, which include specified mole transfers of reactants, kinetically controlled reactions, mixing of solutions, and temperature changes; and (3) inverse modeling, which finds sets of mineral and gas mole transfers that account for differences in composition between waters, within specified compositional uncertainty limits.

PHREEQC can be applied to calculate concentrations of elements, molarities and activities of aqueous species, pH, pe, saturation indices, and mole transfers of phases to achieve equilibrium as a function of specified reversible and irreversible geochemical reactions. In the present study,

PHREEQC was used to calculate the aqueous species of arsenic under different pH, ion-exchange equilibria, and surface-complexation equilibria during the separation and precipitation experiments under different experimental conditions, which may be met for AsO_4^{3-}/AsO_3^{3-} recovery from the reaction solution in lab.

3.3 Preparation of specific solutions

For the experiments in laboratory, all glass flasks and tubes were washed by 1 % HNO_3. All the solutions with different anions were prepared as stock solution before using. Stock solutions of 500 mg/L As(III) and also of 250 mg/L As(V) were prepared by dissolving extra pure reagent sodium arsenite ($NaAsO_2$) and extra pure reagent sodium arsenate ($Na_2HAsO_4·7H_2O$) in 1 L deionized water (18.2 MΩ.cm Milli-Q), respectively. 10 g/L stock solution of Cl^-, SO_4^{2-} and PO_4^{3-} were prepared by dissolving NaCl, K_2SO_4 and KH_2PO_4 in 100 mL of deionized water, respectively. All stock solutions were kept in fridge at 4 °C, except As(III) solution, which was kept under argon atmosphere in a glove box with oxygen content of less than 3 ppm, 1 bar pressure and a constant room temperature of 21±1 °C. The deionized water was refreshed to remove O_2 out by pure N_2 gas for 15 min before using. Tap water, which was from the tap water supply system of Karlsruhe, was applied for the whole selective separation experiments. 65 vol% or 69 vol% HNO_3, 32 vol% NaOH and 28 % NH_4OH were used to adjust the pH to the required values in the different steps of the experiments. 96 % ethanol was used for washing precipitate. pH values were measured by pH indicator or pH meter. For the desorption experiments, eluents with different concentrations were prepared from 32 vol% NaOH, 37 vol% HCl, 96 vol% CH_3COOH, dissolved NH_4NO_3 and $NaNO_3$. The producers of all the chemicals and apparatus are introduced in *Appendix 1.*

3.4 Analytic methods

Many different analytic methods were used in the study. Accurate and reliable methods for the measurement of chemicals like As(III), As(V), P(V), Cl^-, SO_4^{2-}, NO_3^-, HCO_3^-, were required. The concentrations of acid matrix that used for desorption experiment were also needed to be quantified. The methods were needed for the characteristics of different solids, like different adsorbents for the separation experiments, and also the precipitates, which were formed in the experiments.

3.4.1 Concentration of acid

The method of acid-base titration (Hesperides, 2007) is the determination of the concentration of an acid or base by exactly neutralizing the acid or base with an acid or base of known concentration. This allows for quantitative analysis of the concentration of an unknown acid or base solution. It uses the neutralization reaction that occurs between acids and bases and the knowledge of how acids and bases will react if their formulas are known.

In the study, this method was applied for detecting the concentration of HNO_3, which was very important for defining the exact conditions for selective separation of arsenic oxyanions from water. The concentration of 65 vol% and 69 vol% HNO_3 was identified by 0.25 M NaOH standard solution and the indicator Phenolphthalein solution. Two mL of 65 vol% and 69 vol% HNO_3 was exactly pipetted in 20 mL deionized water to obtain a diluted HNO_3 solution, which was mixed with 2-4 drops of 0.2 % Phenolphthalein solution in 100 mL Erlenmeyer flask. Meantime, 50 mL of 0.25 M Standard NaOH was added accurately in the burette. Then the standard NaOH solution was titrated in HNO_3 solution, until the HNO_3 solution is from colorless to pink. The titration process was repeated 3 times to obtain a set of reliable readings, from which an average was taken.

3.4.2 Concentration of As

Several methods of As analysis were employed, including Fluid Injection Atomic Spectrometry (FIAS), High Resolution Inductively Coupled Plasma-Mass Spectrometry (HR-ICP-MS) and Inductively Coupled Plasma Mass Spectrometry-Optical Emission Spectrometry (ICP-OES). The properties of each method are compared as Table 3.1. Low concentrations (< 1 mg/L) of As were analyzed by fluid injection atomic spectrometry (FIAS). Low concentration (< 5 mg/L) of As with other existed elements were simultaneously detected by Inductively Coupled Plasma Mass Spectrometry (ICP-MS). Whereas, high concentrations (>5 mg/L) of As were analyzed by ICP-OES, which has a relatively high detection limit (>0.8 mg/L). For detection of As(III), an alumino-silicate based sorbent (MAS) (Meng & Wang, 1998; Meng et al., 2001a) was additionally used to retain As(V), then As(III) in effluent was detected by FIAS.

Table 3.1: Different methods for As analysis.

Methods	Analysis principle	Range of calibration curve	Average DL (3*Sigma)	Advantage
FIAS	Separating As as gaseous hydride from sample solution (Pavicevic et al., 2000)	0-20 μg/L	0.3 μg/L	Detecting As with a low concentration; Inexpensive
HR-ICP-MS	Ionizing the sample with inductively coupled plasma and then using a mass spectrometer to separate and quantify those ions. (Heinrichs et al., 1998)	0-2500 μg/L	0.02 μg/L	Multi-element analysis of liquid samples; Detecting As with a low concentration
ICP-OES	Using the inductively coupled plasma to produce excited atoms and ions that emit electromagnetic radiation at wavelengths characteristic of a particular element (Mermet, 2005; Stefánsson et al, 2007)	0-100 mg/L	0.8 mg/L	Detecting As with a high concentration

3.4.3 Concentration of phosphate, chloride, nitrate, sulfate and bicarbonate

The main anions in experimental solution without high concentration of HCl or HNO_3, like PO_4^{3-}, Cl^-, NO_3^- and SO_4^{2-}, were measured by ion-chromatography (IC), whose detection limit and calibration range are shown in Table 3.2. For detection of PO_4^{3-} concentration, the detection limit of the ammonium-molybdate method (0.005 mg/L), which is according to DIN EN ISO 6878-D11, was quite lower than that of IC (0.08 mg/L). For the accurate measurement of PO_4^{3-}, the low concentration of PO_4^{3-} was determined colorimetrically by using the ammonium-molybdate method, which can also measure PO_4^{3-} in the solution with much high concentration of acids, like HNO_3 or HCl. In this method, ascorbic acid is used to reduce the molybdophosohate species to molybdenum blue, which has a maximum absorbance at 880 nm by the spectrometry (Clesceri et al., 1998). The concentrations of Cl^- and SO_4^{2-} in a solution with a very low pH, which has too high concentration of NO_3^-, were assessed photometrically, using a field kit with a detection range between 5 and 125 mg/L for Cl^- and 5-250 mg/L for SO_4^{2-}, respectively. The concentration of HCO_3^- was detected with acid capacity to pH 8.2 and pH 4.3.

Table 3.2: Determined elements by IC measured at given time and average IDL (2.764*Sigma, in mg/L) from all monitoring and in-situ experiment analyses (n=10).

Method		F^- (mg/L)	Cl^- (mg/L)	Br^- (mg/L)	NO_3^- (mg/L)	PO_4^{3-} (mg/L)	SO_4^{2-} (mg/L)
IC	Calibration	0-20.00	0-40.00	0-80.00	0-80.00	0-120.00	0-80.00
	Average DL	0.01 ±0.004	0.06 ±0.023	0.03 ±0.011	0.06 ±0.020	0.08 ±0.029	0.04 ±0.016
Photo-metry	Calibration		5-125			0-2.5	5-250
	Average DL		5			0.005	5

3.4.4 Characteristic of adsorbents and precipitates

The element contents of the different adsorbents were quantitatively measured by X-ray fluorescence (XRF). A rhodium X-ray tube was used as radiation source whereas a Si (Li)-detector was used for detection and quantification. In order to optimize the fluorescence measurements each sample was analyzed by consecutively applying an Al-, Cu- and Pd-filter. Precision (better than 5%) was calculated from repeated measurements. The recovery (better than 10 %) was checked by different reference materials, i.a. GXR 2, GXR-3, GXR-4, GXR-5, GXR-6 (Gouveia et al., 1994).

Particle size distribution of adsorbents and also the impression about the composition of the mineral grains and, especially, about their surface topography, which were produced from whole AsO_4^{3-} or AsO_3^{3-} separation procedure, were detected by high solution electric microscopy (EM) and scanning electron microscopy coupled with an X-ray fluorescence detection device (SEM-XRF). The measurement of materials with big particle size was measuremed with EM, and the measurement of materials with small particle size was done by SEM-XRF. The surface areas of each adsorbent were detected by BET (BEL Japan, Inc.), which was done in Engler-Bunte-Institute Division of Combustion technology (EBI-VBI).

XRD analyses of the arsenic precipitate were carried out at the Geological Institute of the University of Lausanne, Switzerland. About 500 mg of sample powders were pressed (20 bars) in a powder holder covered with a blotting paper and analyzed by XRD (Xtra Diffractometer, Thermo, Ecublens, Switzerland) at 45 kV and 40 mA. 1.5406Å CuKα1-radiation was used at angles between 2° and 65°. Identify of arsenic precipitate was analyzed by matching standard reflection from powder diffraction files (the International Centre for Diffraction Data, Newtown Square, PA, USA).

3.4.5 Measurement of composition of oxygen isotope in solid and water phases

3.4.5.1 Analysis of oxygen isotope in solid phase

According to the technique used for the determination of the oxygen isotope ratios in PO_4^{3-} (Kornexl et al., 1999a; LaPorte et al., 2009; Lecuyer et al., 2007; McLaughlin et al., 2004; Vennemann et al., 2002), the oxygen isotope composition in the precipitated silver arsenate was determined by on-line coupling of Element Analyzer (EA) with high temperature reduction with graphite and continuous flow mass spectrometry (IRMS) (EA-Py-CF-IRMS). The systemic detection deviation of $\delta^{18}O$ composition of solid phase in the lab of IMG is ±0.5 ‰.

The comparable methods for evaluation of oxygen isotope were done in University Lasausane. The EA-Py-CF-IRMS method was set up by using the Finnigan MAT Delta^{plus}XL, which is an extended ion optics version of the Delta^{plus} for atmospheric research and the on-line measurement of hydrogen isotope ratios in a He carrier gas stream. The instrument is equipped with an Agilent 6890 GC, with a CTC A200S autosampler, and is coupled to the MS *via* a Finnigan MAT GCCIII interface.

About 0.5-0.8 mg of Ag_3AsO_4 resulted from the conversion of the samples weighted out into silver capsules that are 2mm in diameter and 3 mm high. The silver capsules with samples are tightly folded to minimize the amount of trapped air and loaded into the autosampler that was purged with He gas. Then the folded sample cup was allowed to fall into the ceramic reactor in HEKAtech high temperature oven coupled to an EuroVector element analyzer (TC-EA), which was held at 1450°C to pyrolyze Ag_3AsO_4/Ag_3AsO_3 to CO gas and continuously flushed with He (purity 99.999 %; flow rate about 90 ml/min, 0.7 bar). The ceramic reactor consists of a glassy carbon tube packed with glassy carbon chips and a small graphite reaction cup. After passing the GC column of the EA (hold at 75°C), the generated CO gas is transferred by a continuous He flows into an IsoPrime (GV Instruments) mass spectrometer. Reference CO (9 psi) is introduced through the reference gas box and open split interface of the mass spectrometer.

For detection in samples, the $\delta^{18}O$ values (VSMOW) were then adjusted to the VSMOW scale using the international standards of USGS47 (H_2O), VSMOW (H_2O), IAEA N3 (KNO_3), NBS 127 ($BaSO_4$), Ag_3PO_4 from NBS 120c (internal standard in IMG, KIT, (Vennemann et al., 2002)), internal Ag_3PO_4 standard (LK-2 and LK-3) with light and heavy oxygen isotope, and also the newly developed internal standard materials Ag_3AsO_4 for accuracy control. Each of them was weighted as 0.18-0.20 mg of KNO_3 (corresponding to ca. 80 µg of O), 0.29-0.31 mg of $BaSO_4$, 0.50-0.80 mg of Ag_3PO_4. For measurement of the $\delta^{18}O$ value in arsenic oxyanions, 0.5-0.8 mg of Ag_3AsO_4 was weighted out into silver capsule. The analytical precision was estimated by multiple sample replicates (triplicates), while the accuracy was assessed by analyzing the reference materials Ag_3PO_4 and internal standard Ag_3AsO_4 among every four samples. The $\delta^{18}O$ values were calculated in per mil (‰) difference using Isodata 2.0 software and Eqn. (1):

$$\delta^{18}O = \left(\frac{R_{sample} - R_{ref}}{R_{ref}}\right) 1000 \quad (1)$$

where R_{sample} and R_{ref} are the ratios of $^{18}O/^{16}O$ for the sample and reference gas, respectively. The precise of measurements for each reference material and establishment of calibration are detailed described in Chapter 7.

3.4.5.2 Analysis of oxygen isotope in liquid phase

Oxygen isotope in liquid samples, which was generate during the separation and precipitation experiments, was detected by Liquid water isotope analyzer (LWIA-24d) (Los Gatos Research) and standardized to the VSMOW reference material (Coplen et al., 2002). Sample volume for injection was less than 10 µLm and the high precision (1-sigma) for $^{18}O/^{16}O$ is 0.1 ‰. The sample temperature and operating temperature are 0-50 °C and 5-45 °C, respectively. The systemic detection deviation of $\delta^{18}O$ composition of water in the lab of IMG is ±0.08 ‰. Analysis quality was regularly checked with three certified reference solutions (Table 3.3).

Table 3.3: Precision for isotopic reference materials, measured by LWIA.

Parameter	Reference material	n	Reference value (VSMOW)	Analysis (VSMOW)
$\delta^{18}O$ value	LGR3A	10	-13.08±0.04 ‰	-13.1±0.15 ‰
	LGR4A	10	-7.72±0.08 ‰	-7.69±0.15 ‰
	LGR5A	10	-2.78±0.04 ‰	-2.80± 0.15 ‰

4 Adsorbents selection and their adsorption properties of arsenate and arsenite

4.1 Introduction

The focus of the study is to develop an efficient method for separation of AsO_4^{3-} and AsO_3^{3-} from other common oxygen bearing anions occurring in natural water, like PO_4^{3-}, HCO_3^-, and SO_4^{2-}. Because both AsO_4^{3-} and AsO_3^{3-} can be most conveniently extracted in form of insoluble silver salts, using a $AgNO_3$ solution, the presence of NO_3^- anions is not critical, but on the other hand, the procedure must allow to reduce the concentration of Cl^- ions in order to avoid the formation of large amounts of AgCl in the precipitate. For this aim, two basic rules to select adsorbents are necessary: (i) the adsorbents must be very effective for extracting arsenic from water; (ii) it potentially is possible to separate arsenic from other interfering ions. More than 600 adsorbents have been reported to apply for the removal of arsenic in drinking water, groundwater and wastewater (Mohan & Pittman, 2007). Many techniques used for the removal of arsenic, prior to adsorption, are oxidizing As(III) into As(V), since arsenic acid is dissociated over a far larger pH range than arsenous acid. However, such techniques are inappropriate for the purpose here, because they evidently would alter the isotopic composition of oxygen. A further requirement is the possibility to enrich the relatively low initial AsO_4^{3-} or AsO_3^{3-} concentrations during the separation. In most convenient way, this can be done by loading the arsenic content from a large volumes of water on a sorbent, from which AsO_4^{3-} or AsO_3^{3-} can be stripped within a small volume of eluent. Consequently, a material, which is suitable for the purpose of this study, must has a high adsorption capacity for AsO_4^{3-} or AsO_3^{3-}, but not for other anions, like PO_4^{3-}, HCO_3^-, SO_4^{2-} and Cl^-.

Based on the aim to extract arsenic from water for measuring the oxygen isotope composition in arsenic oxyanions, the adsorbents for the study must be effective and selective for removal of As(V) or As(III) from water. Eight adsorbents, which have been proved as the effective adsorbents to adsorb As(V) or As(III) from water, were selected for the proposal of separation of AsO_4^{3-} or AsO_3^{3-} for the isotopic composition of oxygen: (i) an aluminosilicate based sorbent (MAS) (Meng & Wang, 1998), (ii) IRA-400 (OH) (IRAN), (iii) N-methyl-D-glucamine on poly - (vinylbenzyl chloride) beads (V-NMDG) (Dambies et al., 2004), (iv) porous polymer beads loaded with monoclinic hydrous zirconium oxide (MHZR) (Suzuki et al., 2000), (v) Basic Yttrium Carbonate (BYC) (Wasay et al., 1996), (vi) Fe_2O_3 nano-particle (NFO), (vii) Muromac B-1-La-OH (MBLO) (Trung et al., 2001), and (viii) Analig AS-01 (AA0) (Rahmani et al., 2011). Their preparations, characteristics, and also the results of the test for their adsorption of AsO_4^{3-} and AsO_3^{3-} are presented and discussed in this chapter.

4.2 Characteristic of each selected adsorbent

4.2.1 Meng's cartridge (MAS): an alumino-silicate based adsorbent

Alumino-silicate based adsorbent (MAS), which was developed for field separation of arsenic species to enable the detection of their respective concentrations, is always packed in a cartridge to be used on site to selectively adsorb As(V) from groundwater to measure concentration of As(III) (Meng & Wang, 1998). This popular field material adsorbs As(V) but not As(III). Arsenic(V) can be removed approximately in 50 mL water with the concentration of As(V) was less than 500 µg/L As(V) by 2.5 g MAS (Meng et al., 2001a; 2001b; Guo et al., 2008b). The efficiency of these cartridges in terms of the recovery of the oxyanionic As species was 95-98%. Although any details on the composition of commercial MAS are not published, the EDXRF data shows the qualitative content of different elements in MAS (Table 4.1). Quite high contents of sulfur, silicon, as well as aluminum were detected in MAS. The little peak of titanium and iron was also in the spectrum of XRF. The fate of each element during the further tests will be presented later.

Table 4.1: EDXRF data for original MAS.

Compound	Al_2O_3 (%)	SiO_2 (%)	S (ppm)	Ti (ppm)	Fe_2O_3 (%)
MAS	14	82	22000	270	0.118

4.2.2 Preparation of IRA-400 (OH) (IRAN)

Amberlite IRA-400 is a macroporous strong basic anion exchange resin with a large application field. The resins with different forms (such as OH^-, Cl^-, SO_4^{2-}, Br^-) were often applied to remove anions like NO_3^-, chromate and PO_4^{3-} from solutions (Chabani et al., 2007; Mustafa et al., 1988; 1989; 1990; 1992; 2004; 2009). The adsorption behavior of PO_4^{3-} on different forms of amberlite IRA-400 (Cl^-, Br^-, NO_3^-, and SO_4^{2-}) is almost similar, but the ratio of the exchanged anions (Cl^-, Br^-, NO_3^-, SO_4^{2-}) to the amount of adsorbed PO_4^{3-} increases with the increase in pH and decreases with the concentration of PO_4^{3-} in solution. The reason for this behavior must be sought in the electro-selectivity of the resin. In a pH range between 3 and 11, the ionic species of PO_4^{3-} in solution are $H_2PO_4^-$ (pK_1=2.15), HPO_4^{2-} (pK_2 =7.20) and PO_4^{3-} (pK_3=12.38). From a dilute solution, the resin prefers to adsorb the ion with the highest charge (PO_4^{3-}), consequently the sorption of PO_4^{3-} increases with increasing pH. However, no papers report about the use of IRA-400 for the selective separation of arsenic oxyanions from PO_4^{3-} or other common oxyanions.

The Amberlite IRA-400 (OH) (16-50 mesh) (Sigma-Aldrich), which was applied in this study, is a strong base, anion exchange resin with a quaternary ammonium functional group, with the main characteristics given in Table 4.2. In order to remove trace impurities, before use, the resin was first washed to remove the Cl^- from the surface of resin with de-ionized water (18.2 MΩ.cm MilliQ), and then with 4 L of 0.25 M NaOH flowed until no Cl^- can be detected in the effluent (Figure 4.1). The concentrations of different elements in IRA before and after being washed by NaOH solution were detected by EDXRF (Table 4.3). The results of EDXRF show that most of Cl^- was removed from the surface and inside of IRA (OH) resin, which means OH^- from NaOH solution was instead of most of Cl^-, as well as aluminum in the IRA (OH) resin, which reacted with NaOH as $Al(OH)_3$ and dissolved in NaOH solution. After washing, the resin, which is named here IRAN, was kept in wet and closed bottle for further analyses.

Table 4.2: Characteristics of IRA-400 (OH).

Limit	60 °C max. Temperature
Cross-linkage	8%
Moisture	~45%
Matrix	Styrene-divinylbenzene (gel)
Particle size	16-50 mesh
Operating pH	0 – 14
Capacity	1.2 meq/mL by wetted bed volume

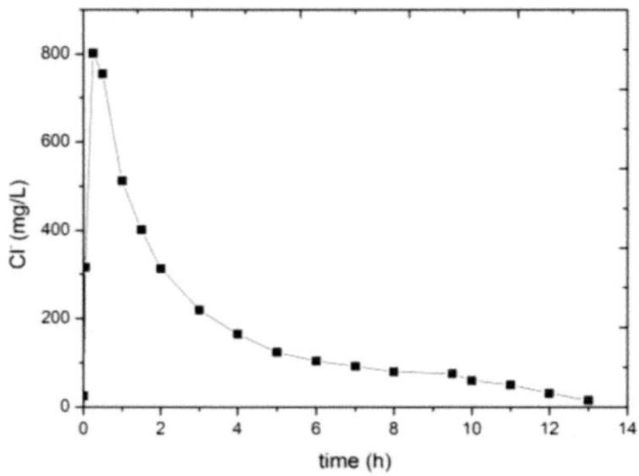

Figure 4.1: Concentration of Cl^- in outflow during procedure of IRAN washing. Reaction condition: Inflow: 4 L of 0.25 M NaOH; Flow rate: 5 mL/min; IRA-400 (OH): ~26 g.

Table 4.3: Comparison of EDXRF data for original IRA and IRAN after washed with 4 L of 0.25 M NaOH. IRA is the material before washing and IRAN is the material after washed.

Compound	Al_2O_3 (%)	Cl (ppm)	K_2O (%)	CaO (%)	Fe_2O_3 (%)	Ti (ppm)
IRA(OH)-Original	1.32	141000	0.02	0.02	0.007	12.4
IRAN	0.13	1630	0.02	0.02	0.002	10.7

4.2.3 Synthesis of N-methyl-D-glucamine on poly - (vinylbenzyl chloride) beads (V-NMDG)

N-methyl-D-glucamine on poly-(vinylbenzyl chloride) beads (V-NMDG) were reported as an AsO_4^{3-}-selective resin for arsenic removal from water (Dambies et al., 2004). Arsenate selectivity depends upon immobilizing the NMDG on a flexible polymer support and having the amine in the protonated form, where the protonated NMDG resin cross-linked with 2 wt% divinylbenzene has a great selective and effective for adsorption of AsO_4^{3-}. V-NMDG selectively adsorbs As(V) when PO_4^{3-}, SO_4^{2-} or Cl^- ions were present in solution at high concentrations and slight acidic pH range (Dambies et al., 2004) .

V-NMDG was synthesized by immobilizing N-Methyl-D-glucamine in polyVBC beads (size 180-250μm), which was with 2 wt % divinylbenzebe Cross-linked (DVB). In the synthesis experiment (Dambies et al., 2004; Zhu & Alexandratos, 2005), 20 g N-Methyl-D-glucamine was dissolved in 10 mL of water and 100 mL of dioxane by heating to reflux in a 250 mL three-neck roundbottomed flask equipped with a condenser, thermometer, and overhead stirrer. 2 g of PolyVBC beads were swollen in 20 mL of dioxane and then added to the hot glucamine solution. The mixture was refluxed for 2 h. The beads were washed with dioxane, aqueous dioxane (1:1), water, and then purified by eluting with 1 L each of 4% NaOH, water, 4% HCl, and water. The final resins were eluted with 1M HNO_3 to protonate the amine moiety and dried in a vacuum oven for 17 h at 60 °C. The structure of the bifunctonal polyol of V-NMDG is showing in Figure 4.2, in which the amine moiety is separated from the polyol group. The load capacity of V-NMDG was 1.77 mmol/g.

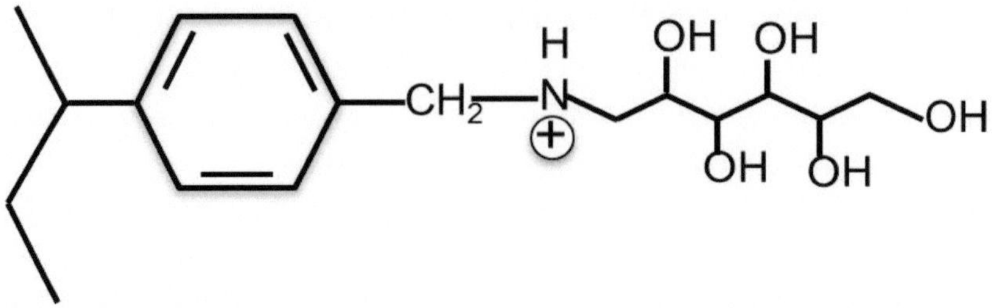

Figure 4.2: Molecular structure of the bifunctional polyol of the V-NMDG compound.

4.2.4 Preparation of the porous polymer beads loaded with monoclinic hydrous zirconium oxide (MHZR)

The porous polymer beads loaded with monoclinic hydrous zirconium oxide (MHZR) was reported as effective adsorbent for AsO_4^{3-}, AsO_3^{3-}, PO_4^{3-}, fluoride, Lead (II) and Titanium dioxide by (Suzuki et al., 1997; 2000; 2001). The presence of common anions including Cl^-, NO_3^-, $CH_3CO_2^-$ and SO_4^{2-} did not interfere with the adsorption of As(V) on MHZR in the concentration range up to 100 times to that of As(V) (Suzuki et al., 2000).

MHZR was prepared by loading monoclinic hydrous zirconium oxide on Amberlite XAD-7 resin (Rohm and Haas Co.), following the procedure outlined in *Suzuki et al.* (2000). The polymer resin, Amberlite XAD-7 (Rohm and Haas Co.) was washed with dioxane in a Soxlhet extractor for 5 hours to remove impurity substances, and then dried in a vacuum desiccator at 50 °C over night. 21.9 g of zirconyloxy-chloride ($ZrOCl_2$•8 H_2O) (Aldrich) was prepared in 250 mL of methanol. To this solution, 17.0 g of dried Amberlite XAD-7 beads were added and the mixture was placed in a vacuum desiccator for 30 min. Methanol was evaporated off for 30 min by using rotary evaporator in order to incorporate zirconium salt into the pores of XAD-7 resin (Borgo & Gushikem, 2002). To the dried residue, 150 mL of 28% ammonia solution was added and the contents were stirred for 5 h to hydrolyze the $ZrOCl_2$ into hydrous ZrO_2. 750 mL of water was added to remove ZrO_2, which was formed out the resin bead, out of the resin beads, and the white precipitates so formed were removed by decantation.

Decantation procedure was repeated to remove the precipitants until the supernatant became clear. Then, the resin was heated with water to crystallize the zirconium species. The formed monoclinic crystals deposited inside the pores by heating the resin in acidic solution (pH=2) at 150 °C for 15 h in an autoclave. These resins thus finally obtained were washed with ethanol, dried under reduced pressure at 50°C, and kept in a desiccator until use. MHZR is expressed as $ZrOH_2^+ X^-$, where X^- stays for the exchangeable anion. The particle size of the loaded resin varied between 60 and 140 µm by measuring electrometric microscopy (EM) and calculating with ImageJ software, whereas its surface area was determined at 373 m^2/g by using BET (Figure 4.3).

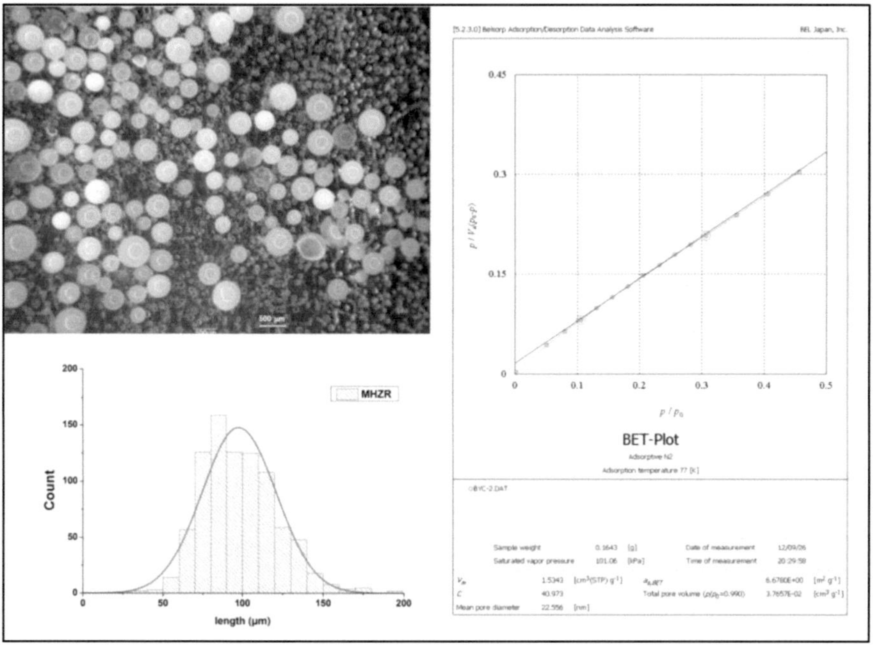

Figure 4.3: EM and BET analysis for the size distribution and surface area of MHZR.

4.2.5 Preparation of Basic Yttrium Carbonate (BYC)

Basic Yttrium Carbonate (BYC) has been reported to be able to remove both of As(III) and As(V) but also PO_4^{3-} under laboratory conditions (Haran et al., 1997; Wasay et al., 1996). Anions such as Cl^-, Br^-, NO_3^- and SO_4^{2-} have no interference in the removal process of As(III) and As(V) (Wasay et al., 1996) in the range of optimal pH value for adsorption.

The preparation of BYC was carried out according to (Sordelet & Akinc, 1988). 12.1 g $YCl_3{\cdot}6H_2O$ (Aldrich 99 % trace metal basis) and 72.7 g Urea (NH_2CONH_2) (GR, Sigma) were added to 800 mL of de-ionized water and mixed for 40 min at 100 rpm (pH = 4.87). Subsequently, while proceeding with the stirring, the solution was heated and kept boiling for another 1 hour. When the temperature reached about 90°C, urea started to decompose and a homogeneous precipitate of white BYC started to appear. Unlike the precipitation of aluminum ions with urea (Blendell et al., 1984; Cornilsen & Reed, 1979; Willard & Tang, 1937), where hydroxide ions generated by decomposition of urea play the critical role, the carbonate ions generated by the urea decomposition control the precipitation of BYC. Urea processes negligible basic properties (K_b=1.5*10^{-14}), is soluble in water, and it rapidly hydrolyses at 90-100 °C. The decomposition of urea into ammonia and carbon dioxide is as the formula (1) (Mendham & Mendham, 2006).

$$NH_2\text{-}CO\text{-}NH_2 + H_2O = 2NH_3 + CO_2 \qquad (1)$$

After the pH increased to 7.0, the white precipitate was washed with deionized water and 96 vol% aceton, until the conductivity of the suspension decreased below 3 µS/cm BYC is expressed as $[Y(OH)]CO_3$. The yield approached an [urea]/[Y^{3+}] ratio of 30, which is an important qualitative indicator for spherical and monosized $[Y(OH)]CO_3$. It is with an optimal recovery and optimal specific surface area of BYC (Sordelet & Akinc, 1988). The particle size of the precipitate was 0.25-0.3 µm by measuring with SEM and calculating with ImageJ, whereas its specific surface area was determined as 6.68 m^2/g by using BET (Figure 4.4).

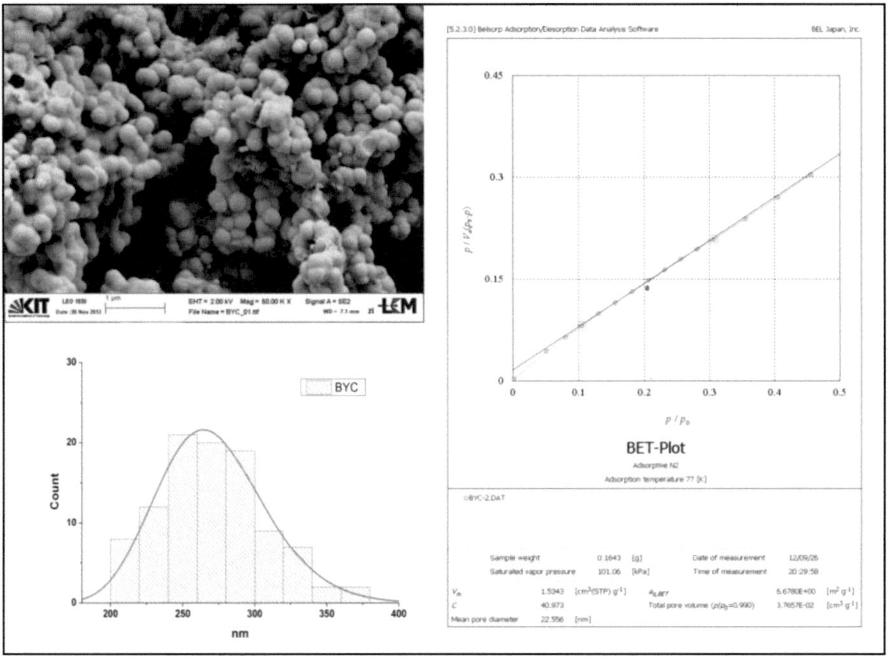

Figure 4.4: SEM and BET analysis for the size distribution and surface area of BYC.

4.2.6 Synthesis of Fe_2O_3 nano-particle (NFO)

The efficiency of the removal of arsenic via ferrihydrite co-precipitation was verified for and proved under different Fe/As ratios, pH values and with different structural characteristics. However, the selectivity of these materials in respect of the other common ions like Cl^- and SO_4^{2-} is not known (Richmond et al., 2004; Tang et al., 2011). The iron oxide used in this study is a nanoparticulate Fe_2O_3 produced by flame synthesis in laboratory of inorganic chemistry in Engler-Bunte-Institut Bereich Verbrennungstechnik (EBI-VBT) in KIT. Its properties for As(V) and As(III) adsorption is presented in this study for the first time.

Iron Oxide (Fe_2O_3) nanoparticles (NFO), which were used for the tests, were synthesized in a low pressure hydrogen/oxygen flat flame burner, using iron pentacarbonyl ($Fe(CO)_5$) (Sigma Aldrich) as precursor and argon as carrier gas. Iron pentacarbonyl was delivered in widely variable concentrations from a pressure and temperature controlled bubbler unit in form of saturated vapor in Ar and introduced after a small mixing chamber below the burner matrix into the flame. In the lean, premixed $H_2/O_2/Ar$ low pressure flame $Fe(CO)_5$ is oxidized to Fe_2O_3. Gas flows were controlled by calibrated thermal mass flow controllers and the pressure of the burner chamber was held constant at 70±1 hPa. SEM analysis documents a NFO with a narrow particle size distribution (26.3–36.7 nm) (Nalcacl et al., 2009, 2010a, 2010b). The surface area of the NFO was 75.9 m^2/g (Figure 4.5).

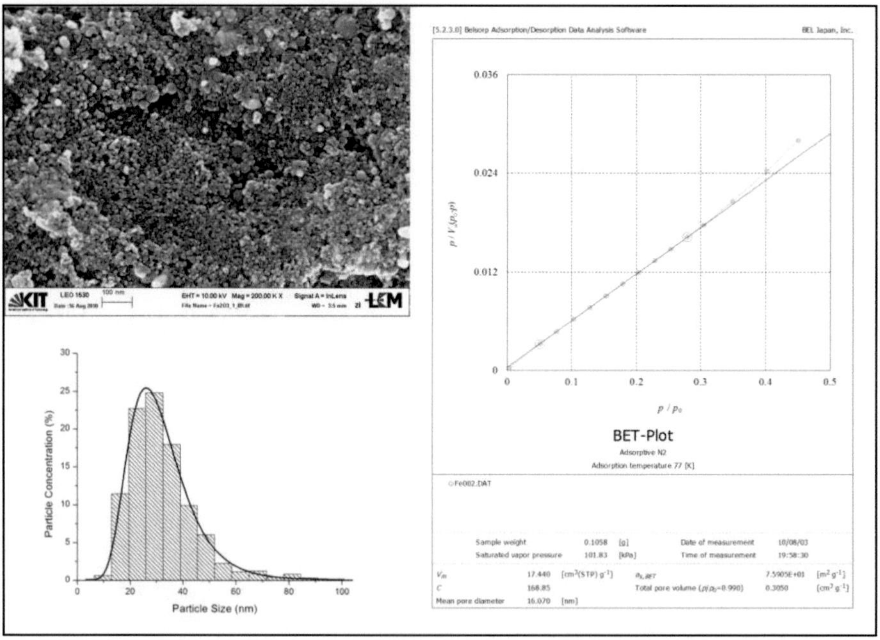

Figure 4.5: SEM and BET analysis for the size distribution and surface area of NFO.

4.2.7 Preparation of Muromac B-1-La-OH (MBLO)

The chelating resin Muromac-A1 in Na^+ form or in Cl^- form (Muromachi Technos Co., Japan) was used to collect As(III) or As(V) for analysis of the arsenic concentration (Jitmanee et al., 2005; Trung et al., 2001). The resin with Cl^- form is not suitable in this study because of too much Cl^- existing in the resin. It was reported that As(III) and As(V) can be quantitatively adsorbed and desorbed from the resin, which loaded La^{3+} on Muromac-A1 in Na^+ form (Trung et al., 2001).

As described in Trung et al. (2001), Muromac B-1–La-OH (MBLO) was prepared by loading La^{3+} in Muromac B-1 chelating resin with Na^+ form (muromachi kgaku Kogyo kaisha, Ltd, Japan), which contains the same compounds in Muromac A1 (Table 4.4). 2.5 g of Muromac B-1 chelating resin Na^+ form was washed with 2 M NaOH and 2 M HCl solutions and then deionized water to remove any trace impurities. Afterwards, the cleaned resin was loaded into the columns and polyethylene frits were fixed on both ends of the columns. An excessive amount of La(III) (4.6 g $LaCl_3.7H_2O$ in 50 mL) solutions of La(III) was fed with a flow rate of 1.0 -1.2 mL/min. The pH of La^{3+} solutions was adjusted to 4 by using sodium acetate buffer. After La(III) was loaded, the column in the metal forms (B1-La) were well washed with 30 mL deionized water to remove any trace metal ions in the resin beds, 5 mL of NaOH solution (3 mM), was passed through the column at a flow rate of 1.5-2.0 mL/min, then 5 mL of deionized water was added to wash the NaOH residue in the resin bed. The metal-loaded chelating resin column in the B1-La-OH form (MBLO) is now ready to adsorb arsenic species from solution (Figure 4.6).

Table 4.4: Typical Physical and Chemical Properties.

Shipping weight (g/L)	795～805
Maximum operating temperature(°C)	80
Total swelling(%) （Na→H）	30
pH range	1.5～9
Ionic form as shipped	Na

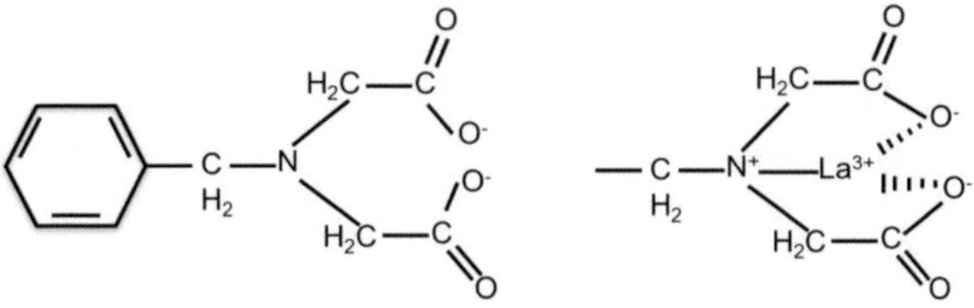

Figure 4.6: Molecular structure of MBLO.

4.2.8 Analig AS-01 (AA0)

Analig As-01 resin (AA0), which is a commercial product, was reported to pre-concentrate As(V) from diluted samples prior to analysis or to separate As(V) from interfering substances (IBC Advanced Technologies, Inc, 2006). Rahman et al. (2011) reported that quantitative adsorption of all arsenic species (As(V), As(III), MMA and DMA) was achieved based on the differences in extraction and recovery behavior of the AnaLig TE-01 (IBC Advanced Technologies, 2007), AnaLig AN-01 Si (IBC Advanced Technologies, 2009) and AnaLig As-01 PA (IBC Advanced Technologies, 2006) columns with pH for different arsenic species. Among those adsorbents, AA0 can separate quantitatively As(V) and As(III) from water samples without interference from the various coexisting ions (Na^+, K^+, Ca^{2+}, Mg^{2+}, Cl^-, NO_3^-, CH^3COO^-, PO_4^{3-}, SO_4^{2-}, ClO_4^-) during this adsorption process. Subsequently, either HNO_3 or a combination of HNO_3 and NaOH were used as eluent to recover the 'captured' species from AA0. Quite high contents of phosphorous, chlorine, and zirconium were detected in AA0 by EDXRF (Table 4.5).

Table 4.5: EDXRF data for the composition of AA0.

Compound	P_2O_5 (%)	Cl (ppm)	Zr (ppm)	Fe_2O_3 (%)
AA0	4.51	3930	13700	0.014

4.3 The first tests for the adsorption properties of arsenate and arsenite

First of all, all the adsorbents, especially the adsorbents, which were produced in IMG lab, were tested for the adsorption of AsO_4^{3-} or AsO_3^{3-}. Afterwards, they were considered to apply for the further experiments. All experimental conditions are specified in the caption of the figures.

4.3.1 Test of Meng's cartridge (MAS)

The first adsorption test by MAS was done in 1000 mL of 100 µg/L As solution containing As(III) and As(V) with a ratio of 4:1 and pH 7. The solution flowed through one cartridge filled with 4 g MAS at the flow rate 60±30 mL/min (Meng et al., 2001a; 2001b). The result is shown in Figure 4.7. After flowing through 1000 ml solution of 100 µg/L As containing As(III) and As(V) with a ratio of 4:1, As(V) was completely adsorbed in MAS, but AsO_3^{3-} stayed practically constant in the effluent.

The adsorption process has been also preformed in 50 mL of tap water containing with different anions and additional As(V), As(III) and PO_4^{3-} (Table 4.6). After shaking with 2.5 g MAS at 150 rpm for 1 hour, As(V) and PO_4^{3-} concentrations decreased below detection limit in the filtrate. Meantime, only a slight decrease was observed in the content of other anions, like Cl^-, NO_3^-, and HCO_3^- but SO_4^{2-} increased much. The effect of the different concentrations of pure As(V) on the adsorption and desorption of As(V) on MAS was investigated in 50 mL solution, with increasing the concentration of As(V) from 0.8 mg/L to 3.2 mg/L, where the desorption of As(V) by using 1 M NaOH solution (Table 4.7). In the As(V) concentration range of 0.8-3.2 mg/L there was no significant differences in the adsorption on MAS, and the average adsorption was 93.8±1.8 %. The first desorption of As(V) from MAS was tested in 50 mL of 0.5 M NaOH for shaking 1 hour at 150 rpm. There was also no big difference among each MAS containing with different amount of As(V), and the average desorption of As(V) from MAS was 83.2±6.4 %. It must be noted that much SO_4^{2-}, around

300 mg/L, was detected in 1 M NaOH in desorption process with 2.5 g MAS. Considering the existed elements in MAS (Table 4.1),_the reason for that high concentration of SO_4^{2-} in solution after desorption process may be that sulfur releases from MAS in the alkaline solution. The results confirm that MAS can be used to separate quantitatively As(III) from PO_4^{3-} and As(V), and also show that As(V) can also be quantitatively leached from MAS. Hence, MAS represents a usable material to work out a method for the further experiments of separation.

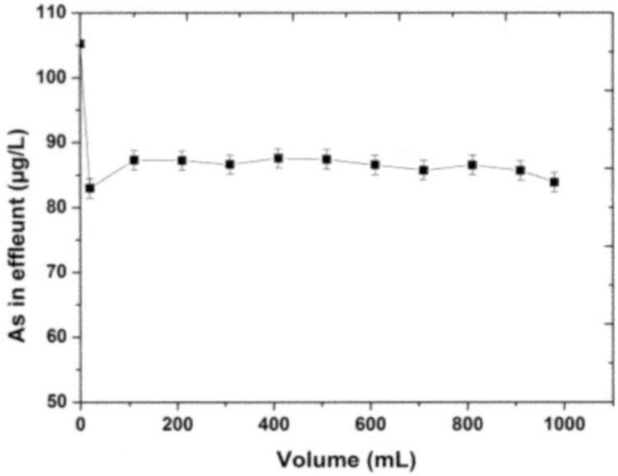

Figure 4.7: Adsorption of As(V) by MAS. Reaction condition: 1000 mL of solution containing 100 µg/L As passed through one cartridge with 4 g MAS, the ratio of As(III)/As(V) was 4:1, pH=7.0, flow rate:50 mL/min, T = 21±1°C.

Table 4.6: Selected chemical parameters in the tap water used for the comprehensive separation experiment, before and after passing it through the alumino-silicate based sorbent (MAS) of Meng & Wang (1998).

Ions in solution	Original solution	Effluent
As(III) (µg/L)	125	125
As(V) (µg/L)	15.3	BDL (<0.33 µg/L)
PO_4^{3-} (mg/L)	5.01	BDL (<0.005 mg/L)
Cl^- (mg/L)	36.0	23.7
SO_4^{2-} (mg/L)	67.8	121
NO_3^- (mg/L)	4.08	4.22
HCO_3^- (mg/L)	342	243

50 mL tap water with 1/10 ratio of As $_{tot}$/P, ¼ ratio of As(V)/As(III) was shaken with 2.5 g MAS at 150 rpm for 1 hour.

Table 4.7: Adsorption and desorption of As(V) on MAS in solution of pure As(V).

As(V) concentration (mg/L)	As(V) adsorbed (%)	As(V) desorbed (%)	SO_4^{2-} in desorption solution (mg/L)
0.8	95.0	77.1	291
1.6	91.7	89.9	311
3.2	94.6	82.6	310

Reaction condition: 1) Adsorption: 50 mL of solution different concentration of pure As(V) passed through one cartridge with 2.5 g MAS, pH=7.0, shaking rate of 150 rpm, T = 21±1°C. 2) Desorption: 2.5 g MAS with adsorbed As(V) put in 50 mL of 1 M NaOH, shaking rate of 150 rpm for 30 min, T = 21±1°C.

4.3.2 Test of IRA-400 (OH) (IRAN)

In order to find out the adsorption properties and the time necessary for a quantitative adsorption of As(V) on IRAN, 1 g of IRAN resin was shaken with 250 mL of solution containing 1 mg/L As(V), and the suspension was stirred at 21 ±1°C for different times (up to 2 hours), using a magnetic stirrer at 300 rpm. These tests showed that IRAN is an effective material to adsorb As(V) quantitatively in less than 1 hour (Figure 4.8).

Solution batches with identical initial 1 mg/L of As(V) and As(III) but different pH (the pH value was 1.5, 7.0 and 11.1) was used in experiments to investigate the effect of protonation on the concomitant sorption of As(V) and As(III) by IRAN (Table 4.8). Under neutral and alkali conditions, both of As(V) and As(III) can be adsorbed, and there was a quite obvious phenomenon that the adsorption of As(V) (98 %-100 %) was stronger than the adsorption of As(III) (86 %-89 %) on IRAN. Whereas, a little of As(V) or As(III) was adsorbed under strong acidic condition (pH=1.5), only 3.5 % of As(V) and 6.0 % of As(III) were adsorbed on IRAN, respectively. The results indicate that IRAN is a possible adsorbent for further experiments testing the separation of As(V) or As(III) from other anions, but not for the separation of As(V) from As(III).

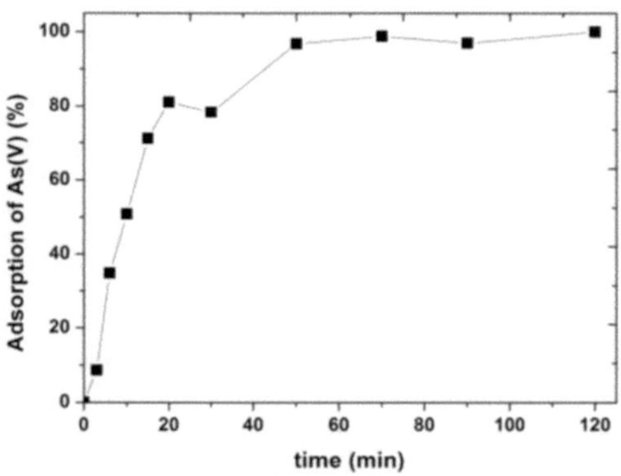

Figure 4.8: Adsorption kinetics of AsO_4^{3-} by IRAN. Reaction conditions: 250 mL of solution with 1 mg/L As(V), 1 g IRAN, pH: 5.5, T = 21±1°C, pH 8.0, stirring at 300 rpm.

Table 4.8: Affect of pH for the adsorption of As(V) and As(III) on IRAN.

pH	1.5	7.2	11.1
As(V) (%)	3.5	98.0	100.0
As(III) (%)	6.0	85.7	89.3

Reaction conditions: 0.5 g IRAN, 250 mL of solution with 1 mg/L As(V) or As(III), stirring 1hour at 300 rpm, T = 21±1°C.

4.3.3 Test of Immobilization of N-methyl-D-glucamine on poly - (vinylbenzyl chloride) beads (V-NMDG)

There are two ionic forms of V-NMDG to be tested, free base and protonated, for the adsorption of As(V) (Figure 4.9). Under the same reaction conditions for 24 hours, 98 % As(V) was adsorbed on V-NMDG in protonated, whereas only 39 % As(V) was adsorbed on the resin in free base form. The results show that AsO_4^{3-} sorption is significantly higher and more rapidly when the ligand is in the protonated form than that in free base form.

The comparison of the adsorption of As(V) and As(III) by V-NMDG resin is shown in Table 4.9. The tests for the concentration of arsenic from 12 to 1200 µg/L were done under same conditions, and pH in solution with As(V) and As(III) was in the ranges of 5.7-6.0 and 9.8-10, respectively, when As(V) and As(III) were charged and the tertiary amine on the resin is protonated. $H_2AsO_4^{3-}$ and $HAsO_4^{2-}$ are the predominate species in solution in a pH range of 4.0-7.2 (Figure 5.1), whereas $H_2AsO_3^-$ is the predominant species for As(III) in solutions at pH >9.1 (Figure 6.1) (Cotton & Wilkinson, 1962). There was almost 100 % (99.7± 0.3%) of As(V) being adsorbed on the resin under different concentration of As(V), but the adsorption of As(III) was always below 10 % (7.8±1.7%). In spite of the weak adsorption of As(III), there was still small amount of As(III) adsorbed, which indicates that V-NMDG can be not applied to separate As(V) from As(III). However, the property for the effective adsorption of As(V) can make it as a worthy adsorbent for the further test of separation of As(V) from other anions in water.

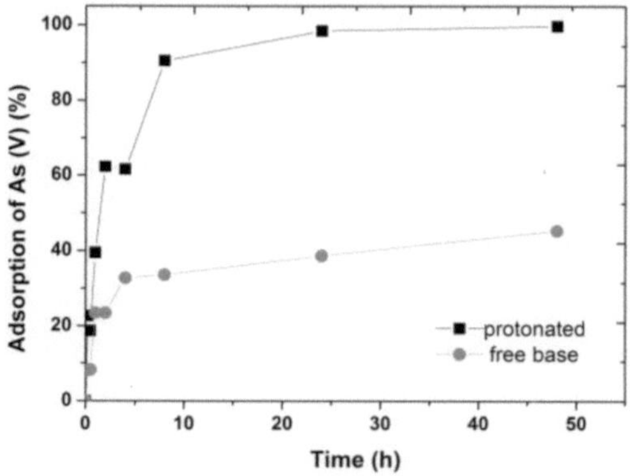

Figure 4.9: Adsorption of As(V) by V-NMDG with different ionic forms. Reaction condition: 1000 mL solution of 1 mg/L As(V), 100 mg V-NMDG, at stirring rate of 200 rpm for 48 hours, pH=6.5, T = 21±1°C.

Table 4.9: Adsorption of As(V) and As(III) under different concentration by protonated V-NMDG.

Initial concentration (µg/L)	**12**	**20**	**100**	**280**	**600**	**1200**
As(V) adsorbed (%)	99.2	99.8	99.9	99.9	99.9	99.9
As(III) adsorbed (%)	8.5	10.2	8.6	7.4	6.7	5.4

Reaction condition: 50 mL solution with As(V) or As(III), 25 mg V-NMDG, and shaking rate: 150 rpm, pH = 5.7-6.0 in As(V) solution, and pH = 9.8-10.0 in As(III) solution, T = 21±1°C.

4.3.4 Test of the porous polymer beads loaded with monoclinic hydrous zirconium oxide (MHZR)

MHZR was first used for checking its adsorptivity of arsenic species. Tests were carried out in 50 mL of solution with different concentrations of As(V) (pH at 7.0-8.0) and As(III) (pH at 9.8-10.2) contacting with 100 mg MHZR for 24 hours at a shaking rate of 150 rpm, according to the method by *Suzuki et al.* (2000). Both As(V) and As(III) were quantitatively adsorbed on MHZR under different concentrations (12-1200 µg/L), showing adsorption rates of 99.4±1.8% and 97.9±1.5% for As(V) and As(III), respectively. The effective adsorption for both arsenic species indicates that preparation of MHZR was successfully and could be applied to the further separation experiments.

MHZR is an effective adsorbent for removal of both As(V) and As(III) in water under different pH values, even more effective for As(V) adsorption. However, co-occurring PO_4^{3-} decreased the adsorption capacity of As on MHZR (Suzuki et al., 2000), which means that PO_4^{3-} can compete with As to complex on the surface of the resin. Therefore, MHZR is proposed to separate As(V) or As(III) from other common anions like Cl^-, NO_3^-, $CH_3CO_2^-$ and SO_4^{2-}, but not from PO_4^{3-}. Prior to these adsorption experiments, preliminary tests were carried out to compare the efficiency of adsorption by percolating the solution through a column versus shaking it with the sorbent (Figure 4.10). Using the optimal flowing rate of 0.55 mL/min, as recommended by Suzuki et al. (2000), As(III) concentration of a solution after passing a column filled with 1 g of MHZR, at the beginning of the experiment, decreased from 119 µg/L to 0.85 µg/L. The concentration in the out-flow increased gradually, and attained 9.79 µg/L at the end of the experiment after ca. 30 hours. The final (average) concentration in the whole liter of solution which passed the column was 6.78 µg/L, indicating that 94.3 % of As(III) was adsorbed on the filling. But, by shaking the same volume (1 L) of solution with the same amount (1 g) of MHZR at 150 rpm, an even higher adsorption rate, of 98.4 %, was achieved within only 3 hours (Suzuki et al., 2000). During this time, concentration of As(III) in solution dropped from 101 µg/L to 1.6 µg/L. The comparison of those two methods shows that the adsorption of As was much faster with respect to the shaking experiment relative to the column experiment.

Table 4.10: Adsorption of As(V) and As(III) under different concentrations by MHZR.

Initial concentration (µg/L)	**12**	**20**	**100**	**280**	**600**	**1200**
As(V) adsorbed (%)	97.8	99.4	99.7	99.8	99.9	99.7
As(III) adsorbed (%)	95.6	98.1	96.8	98.6	99.1	99.4

Experimental conditions: 50 mL of solution contacted with 100 mg of MHZR for 24 hours at a shaking rate of 150 rpm; pH for As(V) solution:7.0-8.0, pH for As(III) solution: 9.8-10.2; T = 21±1°C.

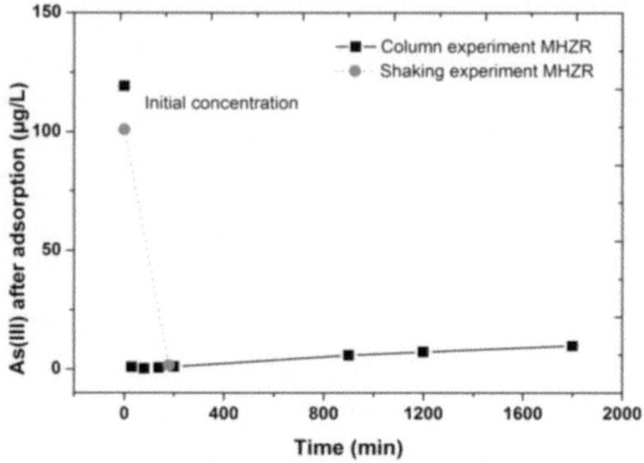

Figure 4.10: Comparison of As concentration after adsorption between column test and shaking tests with HMZR. 1L of solution with 100 µg/L As(III) contacted with 1 g of MHZR; pH: 9.8-10.2. T = 21±1°C. Column Experiment: 1g of dry MHZR was loaded in 3 mL Empty REV column (supelco) and polyethylene frits were fixed on both ends of the column. pH value in 500 mL solution with 200 µg/L As(III) was adjusted from 7.76 to 9.76 by 25 µL 3.2 vol% NaOH. The mixed solution flowed throw the column at a flow rate of 0.85 mL/min. Shaking experiment: Use sealing membrane to wrap bottle to aviod air go inside. The mixed solutions with pH = 9.76 were shaken for 3 hour at a rate of 150 rpm.

4.3.5 Test of Basic Yttrium Carbonate (BYC)

Like tests with MHZR, the experiments for the adsorption of As(V) and As(III) on BYC were carried out in 50 mL of solution with different concentrations of As(V) (pH at 7.0-8.0) and As(III) (pH at 9.8-10.2) contacting with 25 mg MHZR for 24 hours at a shaking rate of 150 rpm. The results show the high adsorption of As(V) and As(III) on BYC which adsorbed As(V) (97.7±0.8 %) more effective than As(III) (91.4±4.7 %) under different concentrations. Although BYC started to be dissolved when pH < 6.5 and the optimal pH values for adsorption of As(V) and As(III) were different, BYC can adsorb also little As(III) (14 %) under the optimal pH range for As(V) adsorption, whereas 85-100 % of As(V) was adsorbed on BYC at the range of pH in 9.0-10.0. Like MHZR, BYC was an adsorbent, which removed arsenic species and PO_4^{3-} simultaneously (Haran et al., 1997; Wasay et al., 1996), therefore, it was also used for the further tests of As(III) separation.

Subsequently, the comparative tests for the efficiency of adsorption on BYC were done by a column and shaking experiment. 1 L of solution with a similar composition was passed with an optimum flow rate (Wasay et al., 1996) of 0.85 mL/min through a column filled with 0.25 g of BYC. The initial As(III) concentration of 99.4 µg/L decreased to 28.4 µg/L at the very beginning of the experiment, indicating that at the given As concentration the used amount of BYC was not sufficient to determine the breakthrough point. During the rest of the experiment, As(III) concentration in the outflow increased gradually, attaining 93.6 µg/L after 20 hours (when 1 L of solution passed the column). The average concentration of As(III) in the 1 L of solution which interacted with the adsorbent was 77.9 µg/L, corresponding to a sorption efficiency of only 21.6 %. In contrast, when the same amount (0.25 g) of BYC was shaken with the same volume of solution (1 L) at 150 rpm for 15 hours (Wasay et al., 1996), concentration of As(III) decreased from 101 µg/L to 8.7 mg/L, (91.4 % of As adsorbed) (Figure 4.11). Consequently, shaking is a much more efficient separation technique than the use of columns when the adsorbent is BYC.

Table 4.11: Adsorption of As(V) and As(III) under different concentrations by BYC.

Initial concentration (µg/L)	12	20	100	280	600	1200
As(V) adsorbed (%)	96.9	97.8	96.8	98.1	97.9	98.9
As(III) adsorbed (%)	93.6	96.6	83.1	91.4	89.6	94.3

Experimental conditions: 50 mL of solution contacted with 25 mg of BYC for 24 hours at a shaking rate of 150 rpm; pH for As(V):7.0-8.0, pH for As(III): 9.8-10.2; T = 21±1°C.

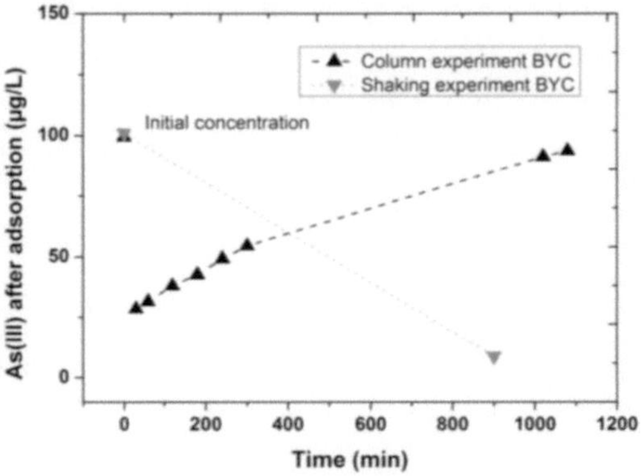

Figure 4.11: Comparison of As concentration after adsorption between column test and shaking tests with BYC. 1L of solution with 100 µg/L As(III) contacted with 0.25 g of BYC; pH: 9.8-10.2. T = 21±1°C. Column Experiment: 0.25 g of dry BYC was loaded in 3 mL Empty REV column (supelco) and polyethylene frits were fixed on both ends of the column. pH value in 500 mL solution with 200 µg/L As(III) was adjusted from 7.76 to 9.76 by 25 µL 3.2 vol% NaOH. The mixed solution flowed throw the column at a flow rate of 0.85 mL/min. Shaking experiment: Use sealing membrane to wrap bottle to aviod air go inside. The mixed solutions with pH 9.76 were shaken for 15 hour at a rate of 150 rpm.

4.3.6 Test of Fe_2O_3 nano-particle (NFO)

As a new material with a small grain size of 26.3–36.7 nm, tests with NFO were only carried out with shaking technique. The adsorption of As(V) and As(III) on NFO was tested in 50 mL of solution with different concentration of As(V) and As(III) at pH range of 7.0-8.0, which was shaking with 10 mg NFO for 24 hours at 150 rpm. Results show that NFO adsorbs both, As(V) and As(III), when pH range at 7.0-8.0. The adsorption of As(V) is more efficient than As(III) under the same conditions (Figure 4.12). The adsorption of each arsenic species decreased slightly as following the increase of concentration of arsenic. It indicates that NFO can be applied for the further separation experiments, and adsorption capacity as well as other adsorption parameters will be clear.

Table 4.12: Adsorption of As(V) and As(III) under different concentrations by NFO.

Initial concentration (μg/L)	12	20	100	280	600	1200
As(V) adsorbed (%)	98.6	99.3	99.6	99.8	99.6	87.1
As(III) adsorbed (%)	97.1	94.0	95.3	94.2	90.1	84.4

Experimental conditions: 50 mL of solution contacted with 10 mg of NFO for 24 hours at a shaking rate of 150 rpm; pH for both of As(V) and As(III):7.0-8.0; T = 21±1°C.

4.3.7 Test of Muromac B-1-La-OH (MBLO)

The experimental procedures for testing adsorption of As(V) and As(III) on MBLO were exactly carried out according to the description in *Trung et al.* (2001). However, in the first two tests, As(III) concentrations monitored in effluent solution kept constant and was the same as that in inflow during the process pasting low concentration (≤ 10 μg/L) of As(III) (Figure 4.12), which indicates that MBLO can not be used to remove or extract As(III) from water.

The possible reason of no adsorption of As(III) on MBLO from the first tests, may be the failure of the MBLO preparation. Therefore, after removing the trace impurities, another MBLO was prepared by stirring 5 g wet weight of Muromac B-1 in 100 mL solution of 0.5 mol/L La at 100 rpm over night instead of passing the La solution through the resin. Then the MBLO resins were washed as the same way as the first preparation process. The new resins were used to adsorb As(V) and As(III) in 50 mL solution with or without other anions (Table 4.13). The adsorption of As(V) or As(III) was still below 10 % in different solutions, and also there was only little adsorption of other coexisted anions like PO_4^{3-} or Cl^-. Those negative results mean that MBLO was prepared in our study could not repeat the results described in the respective paper (Trung et al., 2001), and was not suitable for the proposal of the study.

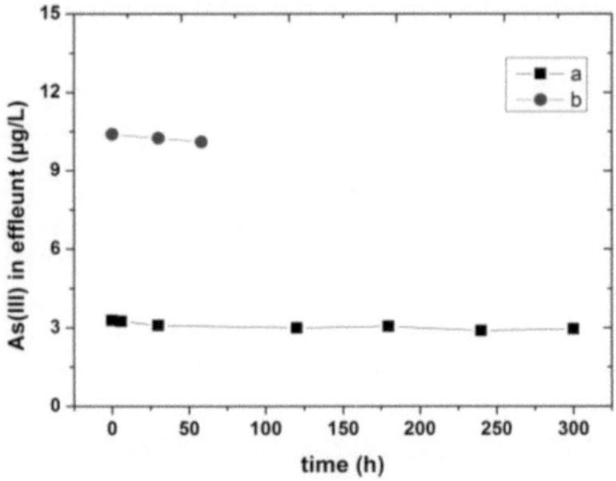

Figure 4.12: As(III) concentration during the process of adsorption of As(III) by column with MBLO. 250 mL of the solution with As(III) flowed through the column with MBLO resins column by flow rate 0.8-1.0 mL/min. MBLO: 1 g. a: solution with 3 µg/L As(III), pH: 5.97; b: solution with 10 µg/L As(III), pH: 6.72.

Table 4.13: The adsorption of each anion in different solutions on MBLO.

Test batch	As adsorbed (%)	Cl^- adsorbed (%)	PO_4^{3-} adsorbed (%)
1.	7.7	-	-
2.	4.4	-	-
3.	5.5	-	13.2
4.	1.9	2.0	-

Experimental conditions: 50 mL of solution with 500 µg/L As(V) or As(III) and different ions, MBLO: 0.5g; reaction time: 2hours, a shaking rate: 150 rpm; pH: 6.35, adjusted by 1M $NaOOCCH_3$; batch 1: solution with pure As(V); 2: solution with pure As(III); 3: solution with As(V)+PO_4^{3-}; 4: solution with As(III)+Cl^-.

4.3.8 Test of Analig AS-01 (AA0)

AA0 was reported as a effective material for removal of As(V) and As(III) without interference of many coexisting ions especially like Cl^-, NO_3^-, CH_3COO^-, PO_4^{3-}, SO_4^{2-}, and the sorption ability of AA0 is based on the molecular recognition and macrocyclic chemistry (Rahmani et al., 2011). Because of the effective property of AA0, the proposal of using AA0 is to separate As(III) from other coexisting ions in water. As the description of As-01 PA, the binding capacity of the resin is approx. 1 mmol/g AnaLig, and the optimal flow rate is 0.25 mL/g/min or less (IBC Advanced Technologies, Inc.), and pH range at 7-9 for As(III) removal (Rahmani et al., 2011). The experiments for adsorption of As(III) on AA0 were exactly under the optimal conditions. The results of two batches are shown in Table 4.14. Arsenite was quantitatively adsorbed on AA0 when the solution contained only 250 μg/L As(III) (Table 4.13.a) or the solution with 250 μg/L As(III) and also 100 mg/L Cl^- (Table 4.12.b) Cl^-. However, much Cl^- was detected in the solutions after adsorption and desorption process in both batches. The released Cl most possibly originated from the AA0 resin, which contained much Cl inside (Table4.5). Since high concentration of Cl still existed in solution after several steps of tests, AA0 was not a suitable adsorbent for extracting As(III) from water.

Table 4.14: The adsorption of As(III) on AA0 Resin under different conditions.

Test batches	**AA0 (g)**	**As(III) adsorbed (%)**	**As(III) desorbed (%)**	**Cl^- released after adsorption (mg)**	**Cl^- released after desorption (mg)**
a	0.50	100.0	37.4	3.84	3.04
b	0.25	97.0	91.9	0.75	1.32

Batch a: 1) Adsorption: 25 mL of solution with 250 μg/L pure As(III) passed through AA0 at a flow rate of 0.2 mL/min, pH: 7-9, T = 21±1°C. 2) Desorption solutions: 1 mL of 0.1 M HNO_3 + 1 mL of 2.0 M NaOH + 2 mL of Mill-Q water. Batch b: 1) Adsorption: 25 mL of solution with 250 μg/L As(III) and 100 mg/L Cl^- passed through AA0 at a flow rate of 0.2 mL/min, pH: 7-9, T = 21±1°C. 2) Desorption solutions: 5 mL of 0.1 M HNO_3 + 5 mL of 2 M NaOH + 10 mL of Mill-Q water.

4.4 Discussion

First of all, As(V) and As(III) in water can be separated by MAS, which can completely adsorb As(V) and PO_4^{3-}, as well as slight amount of other anions, like Cl^-, NO_3^-, and HCO_3^- but not As(III). A small amount of Cl^- and NO_3^- in water did not affect the further experiments of oxygen isotopic measurement in arsenic oxyanions. Meantime, HCO_3^- is easily removed by adjusting pH (Garrels & Christ, 1965). Therefore, the further tests for extracting As(V) from water is focused on separating As(V) from PO_4^{3-} and from a high amount of SO_4^{2-} after being desorbed from MAS by using 0.5 M NaOH. The adsorbents, IRAN, V-NMDG and NFO, which were proved to effectively adsorb As(V), can be applied to separate As(V) from PO_4^{3-} and SO_4^{2-}. The results are presented in chapter 5. For the separation of As(III) from the filtrate, which passes MAS containing As(III), Cl^-, SO_4^{2-} and HCO_3^-, BYC, MHZR and NFO, those three materials have been considered. HCO_3^- needs also not to be taken into account. The fate of Cl^- and SO_4^{2-} during the separation of As(III) is clarified in Chapter 6. The another two adsorbents, MBLO, which does not adsorb As species under the optimal reaction condition as described in *Trung et al.* (2001), and AAO, from which much Cl^- released companying As(III) during the adsorption process, were not applied for the future experiments. Comparison of the efficiency of adsorption by percolating the solution through a column versus shaking, the adsorbents with the big grain sizes (eg. MHZR) is much better than that with a small grain sizes (eg. BYC) to adsorb arsenic in column experiment. However, they showed a similar and high adsorption behavior in the shaking experiments. Although quite different time was required for a near-quantitative adsorption for MHZR, shaking has proven to be a much more efficient separation technique, than the use of columns. A possible reason for the different response of MHZR and BYC to shaking and percolation are the varying shape and size of the particles. MHZR is composed of quite large (60-140 μm) spherical grains so that the solution easily percolates through the inter-granular space, allowing the $H_2AsO_3^-$ ions to interact with the exchangeable X^- on the surface of MHZR. In contrast, BYC is a fine powder with very small grains of only 0.25-0.3 μm in size. The material gets compacted already after a relative small volume of solution flows through the column bed, canalizing the solution along preferential flow paths (e.g., along the wall of the column). Hence, due to the reduced contact times, the exchange process is strongly hindered. Consequently, column experiments are not suitable for materials with small particle size, and all the experiments in the study were performed under shaking conditions. Short summary for the rudimentary experimental parameters of each tested adsorbent for the further tests is presented in Table 4.15. The optimal reaction conditions with different adsorbents for the extraction of AsO_4^{3-} and AsO_3^{3-} from water will be clarified in the chapter 5 and 6.

Table 4.15: The test plan for the further experiments. All experiments were under room temperature (21 ±1 °C).

Adsorbents	For the further experiments	pH	Probably Equilibrium time (hour)	Reaction technique
MAS	Separation of As(V) and PO_4^{3-} from As(III), SO_4^{2-}, Cl^- and NO_3^-, HCO_3^-	4-9	0.5	Shaking or column
IRAN	Possibility for separation of As(V) or As(III) from other existed anions	> 3 for As(V)	1 for As(V)	Shaking
		> 3 for As(III)	1 for As(III)	
V-NMDG	Possibility for separation of As(V) from PO_4^{3-}, SO_4^{2-} and Cl^-	4-6.5	24	Shaking
BYC	Possibility for separation of As(V) or As(III) from SO_4^{2-} and Cl^-	7.5-9.0 for As(V)	1 hour for As(V)	Shaking
		9.8-10.5 for As(III)	15 hour for As(III)	
MHZR	Possibility for separation of As(V) or As(III) from SO_4^{2-} and Cl^-	4-6 for As(V)	2 hours for As(V)	Shaking
		9-10 for As(III)	3 hours for As(III)	
NFO	Possibility for separation of As(V) or As(III) from SO_4^{2-} and Cl^-	?	?	Shaking
MBLO	Not suitable for further experiment			
AAO	Not suitable for further experiment			

5 Separation of arsenate from water

5.1 Introduction

Inorganic oxyanions of trivalent AsO_3^{3-} and pentavalent AsO_4^{3-} are the predominant forms in natural water (Smedley & Kinniburgh, 2002). In most groundwaters, around 20 % As(V) exists in the total arsenic (Guo et al., 2008b). In spite of the little amount of AsO_4^{3-} in groundwater, the oxygen isotope composition in AsO_4^{3-} is still a very important tracer for the study, to provide some vital information of the source of arsenic in the natural groundwater. The tests of different adsorbents for the separation of AsO_4^{3-} from other aqueous oxyanions in water, which were selected before, are presented and discussed in this chapter.

Previously, it was demonstrated that MAS can be used for an efficient separation of AsO_4^{3-} from AsO_3^{3-}, and also other oxyanionic species, such as PO_4^{3-} being co-adsorbed with AsO_4^{3-}, whereas NO_3^- and SO_4^{2-} as well as Cl^- passed together with AsO_3^{3-} and were practically unhindered in the cartridge. Although the producer of MAS does not provide any details on the composition of the sorbent, the separation is based obviously on the property of AsO_4^{3-} and PO_4^{3-} to form strong inner-spheric complexes on aluminosilicates (Violante et al., 2002). The adsorption capacity of AsO_4^{3-} and the state of PO_4^{3-}, and the optimal conditions for separation of AsO_4^{3-} from MAS during the adsorption and desorption process, were also not clear for the aim of the study.

After separation from AsO_3^{3-}, both AsO_4^{3-} and PO_4^{3-} adsorbed on MAS. The key issue in the next step is how to separate AsO_4^{3-} from PO_4^{3-}. Arsenate and PO_4^{3-} have a very close similar hydrolysis behavior (Figure 5.1). Both of them have fairly similar pK_a values (2.0 7.0 and 11.5 for H_3AsO_4 and

2.15, 7.20, 12.38 for H_3PO_4) (Linge, 2002) and analogous chemistry in homogeneous solution (Gresser et al., 1986). Therefore, the quantitative separation of these two anions is a more challenging task, if the initial isotopic composition of the associated oxygen atoms has to be preserved during the procedure. Several methods (Naricise et al., 2005; Tuzen et al., 2009; Xiong et al., 2008) were developed to separate AsO_4^{3-} and AsO_3^{3-} for on-line measurement without influences of the foreign ions, like SO_4^{2-}, PO_4^{3-}, NO_4^{3-}, HCO_3^- Cl^-, but no report compared the concentration change of those oxyanions during the experiment processes. Ross et al. (1961) used a butanol-chloroform mixture to separate PO_4^{3-} from AsO_4^{3-} and got up to 96% PO_4^{3-} extracted, but the extract contained also 13% of AsO_4^{3-} present initial in solution.

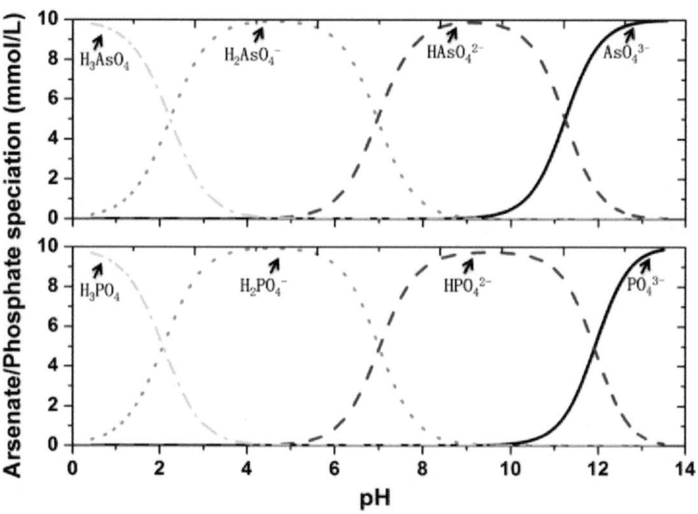

Figure 5.1: Arsenate and PO_4^{3-} speciation as a function of pH for total As(V) and total P(V) concentration 10mmol/L, respectively (by PHREEQC program with water4f.dat and llnl.dat, respectively).

Because even in high arsenic groundwater total As concentrations do not exceed 1-2 mg/L and the As(III)/As(V) ratios are usually high, in addition to separation from other oxyanionic species, it appears mandatory that a proposed procedure allows a pre-concentration of AsO_4^{3-} prior to its precipitation as Ag-arsenate. The basic idea is to load an adequate amount of AsO_4^{3-} onto the cartridge proposed by MAS by passing a sufficiently large volume of groundwater through the cartridge. After leaching the anions from the column, AsO_4^{3-} and PO_4^{3-} are separated by means of IRAN or V-NMDG. It looks that much SO_4^{2-} can release from MAS after desorption of AsO_4^{3-}, which adsorbed on MAS (Table 4.7). MHZR or BYC will be applied to test for the separation of AsO_4^{3-} from SO_4^{2-}. The main aims of this chapter are (1) to clarify the optimal conditions for the adsorption and desorption of AsO_4^{3-} by MAS; (2) to test the selected adsorbents for separation AsO_4^{3-} from PO_4^{3-} and SO_4^{2-}, and find the optimal conditions for quantitative separation of AsO_4^{3-} from water.

5.2 Experiments

5.2.1 Experiment for Adsorption isotherm

The adsorption isotherms were obtained by the batch method employing a constant weight of different adsorbents and 50 mL of the As(V) solution at different concentrations (0.01 mg/L, 0.05 mg/L, 0.1 mg/L, 1 mg/L, 5 mg/L, 10 mg/L, 50 mg/L, 100 mg/L, 250 mg/L, 500 mg/L and 1500 mg/L) of AsO_4^{3-} with fixed amounts of adsorbents in 50 mL bottles for 24 hours at 150 rpm and room temperature (21±1°C). According to the conditions being tested by other authors (Dambies et al., 2004; Meng & Wang, 1998) and also in this study, table 5.1 shows the optimal experimental conditions for the adsorption capacity of the adsorbents, which were mainly applied for the separation of AsO_4^{3-}. The optimum pH of those solutions containing AsO_4^{3-} with each adsorbent was adjusted by using 65 % pure HNO_3 and 28 % conc. NH_4OH.

Table 5.1: Reaction conditions of adsorption isotherm experiments for adsorbents. T: 21 ±1 °C, Reaction time: 24 hours.

Anion	Adsor-bents	Weight (g)	Volume (mL)	pH (experi-mental value)	pH (optimum range)	Shaking rate (rpm)
As(V)	MAS	0.5	50	5.79	4-9	150
	IRAN	1.0	50	5.55	3-11	150
	V-NMDG	0.025	50	5.70	4-6.5	150

5.2.2 Separation experiments

Experiments were carried out with different adsorbents under different pH values, ratios of As/P, reaction times and other parameters, aiming to find the optimal conditions for adsorption of AsO_4^{3-} for separation from other co-occurring oxyanions. An As/P ratio of 1/10, which is typical for many sites with high As groundwater (Charlet et al., 2007; Mukherjee et al., 2009), was applied to many adsorption experiments in the study. The pH of tested solutions was adjusted by HNO_3, CH_3COOH and NH_4OH, where 3.8 is the original pH value in solution without adjusting.

Experiments aiming to strip the adsorbed AsO_4^{3-} from adsorbents were worked out by using different concentrations of acidic or alkaline solution, like HCl, HNO_3, CH_3COOH, NH_4OH or NaOH. All experiments were performed at room temperature (21±1°C). The detailed experimental conditions are specified in the results part and in the caption of the figures or tables. Each experiment was repeated at least three times to prove and improve the precise and correction of the experimental results.

5.3 Results and discussion

5.3.1 Adsorption isotherms of As(V)

The adsorption constants, adsorption densities and adsorption maxima of As(V) for the adsorbents, were evaluated by the isotherm models of Langmuir (1916) and Freundlich (1906) (Figure 5.2), which have been described in section 3.2.1 . The values of the parameters of the Langmuir and Feundlich isotherms, which are calculated from Figure 5.2, are shown in Table 5.2. The correlation coefficients (r^2) for both models range between 0.930-0.999, indicating an excellent fit with the observed data. The Langmuir model, with r^2 >0.990 for all three materials, appears to describe the adsorption slightly better than the Freundlich model.

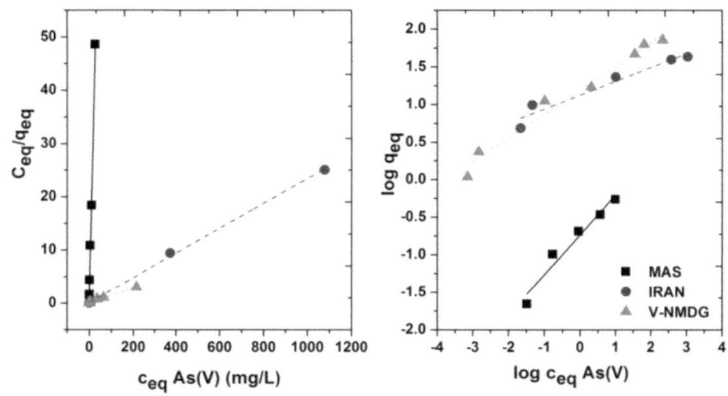

Figure 5.2: As(V) adsorption isotherms on MAS, IRAN and V-NMDG. a: Langmuir isotherm; b: Freundlich isotherm. Experimental conditions: 50 mL batches, with As(V) concentrations in the range between 0.01 mg/L and 1000 mg/L. Shaking rate: 150 rpm; duration: 24 hours. MAS: 500 mg, pH: 7.5; IRAN: 500 mg, pH: 7.5; V-NMDG: 25 mg, pH: 7.0-8.0.

Table 5.2: Isotherm parameters for adsorption of As(V) on MAS, IRAN and V-NMDG.

Materials	Langmiur			Freundlich		
	r^2	q m (mg/g)	k_l (L/g)	r^2	n	k_f ((mg/g)(L/mg)1/n)
MAS	0.993	0.563	0.831	0.943	1.89	0.48
IRAN	0.999	42.9	0.181	0.93	5.46	3.07
V-NMDG	0.991	71.3	0.327	0.973	3.17	3.28

The adsorption maxima of As(V) by MAS, IRAN and V-NMDG estimated with the Langmuir isotherm are 0.563 mg, 42.9 mg/g and 71.3 mg/g, respectively. The fitting curves also suggest that As(V) was adsorbed in form of a monolayer coverage on the surface of these materials. In comparison with the adsorption capacity of As(V) by MAS, which was relative low, only 0.563 mg/L, IRAN and V-NMDG possess a significantly higher value. The Freundlich isotherm applies to both multilayer sorption and non-ideal sorption on heterogeneous surfaces. The close fit of the experimental data with the Freundlich isotherm suggests the heterogeneity of the sorbent surface. According to these data, all three materials appear to efficiently adsorb As(V) in the considered concentration range, especially IRAN (Akpomie et al., 2012; Teng & Hsieh, 1998).

5.3.2 Separation of arsenate by MAS

5.3.2.1 The adsorption of arsenate and phosphate on MAS

The fate of PO_4^{3-} during the procedure of the selective adsorption of AsO_4^{3-} onto MAS was not reported until now. The preliminary results show that at the recommended pH value (pH: 4-9) (Meng & Wang, 1998) not only AsO_4^{3-}, but also PO_4^{3-} practically was completely retained on the sorbent. With a solution containing As(III) and As(V) in a ratio of 1:1, total As content decreased in the effluent to ½ of the initial concentration, while PO_4^{3-} was practically absent, indicating that both PO_4^{3-} and AsO_4^{3-} were retained, while As(III) passed quantitatively through the column with 4.0 g MAS (Table 5.3). By percolating a solution containing $PO_4^{3}$and AsO_3^{3} only, the arsenic content in the effluent remained unchanged (103 and 109 µg/L for each), while PO_4^{3-} was completely adsorbed in the column (decreased from an initial 104 µg/L to 3.9 µg/L in effluent). A solution containing only AsO_4^{3} and PO_4^{3} showed practically no As and P concentrations after passing the column.

Table 5.3: Concentrations of As(III)+ As(V) and P(V) after passing the cartridge with 4.0 g of MAS. 50 mL of solution with pH value of 7.0, and flow rate: 50 mL/min.

Species present in solution		P(V) (μg/L)	As(III)+As(V) (μg/L)
As(III)/As(V)~1/1, P(V)	Initial	105	233
	In effluent	6.1	114
As(III), P(V)	Initial	104	112
	In effluent	2.9	103
A (V), P(V)	Initial	105	109
	In effluent	3.2	2.2

Table 5.4: Concentrations of As and P before and after passing solutions with different As(total)/P ratios through MAS. Initial As(V)/As(III) = ~1:4 (80% As(III)). 50 mL of solution with pH value of 7.0, and flow rate: 50 mL/min. One cartridge contained 2.5 g MAS.

As_{total}/P	Solution	P (μg/L)	% of P Passed	As(V)+As(III) (μg/L)	% As passed
	Blank	1.0		0.3	
1:5	Initial	651		149	
	in effluent	33.1	5.08	121	81.2
1:10	Initial	1320		155	
	in effluent	73.4	5.56	117	75.5
1:50	Initial	7050		165	
	in 1st effluent	563	8.00	128	77.6
	in 2nd effluent	82.3	1.17	121	73.3

In groundwaters PO_4^{3-} is usually much in excess relative to AsO_4^{3-}/AsO_3^{3-}. In order to test the efficiency of the cartridge in the presence of higher PO_4^{3-} concentrations, experiments were carried out in solutions with different As(III+V)/P ratios. Results show that PO_4^{3-} concentrations in effluent increased slightly with decreasing initial As_{total}/P ratios from 5.08 % (at a ratio of 1:5), through 5.56 % (at 1:10) to 8.00 % (at 1:50) (Table 5.4). This is not satisfactory, but passing the effluent for a second time through the column, total P concentration decreased to 1.17 % of the initial value (from 7050 ppb to 82 ppb), even when P was 50 times higher than As. The portion of As which concurrently passed through the column (75.5 and 81.2 %) scatters slightly around the percentage of As(III) in the initial solution (80% at an As(V)/As(III) ratio of 1:4).

In a further experiment, the adsorption capacity of PO_4^{3-} in one cartridge with 2.5 g MAS in treating larger volumes of water was tested (Table 5.5). 1 L of water containing PO_4^{3-} and AsO_3^{3-} was passed a cartridge with 2.5 g MAS at a flow rate of 50 mL/min. In the first 100 mL effluent, the concentration of PO_4^{3-} decreased to 1.7 % of the initial value, while As(III) passed unhindered through the column (Table 5.5). After percolating 900 ml water, the PO_4^{3-} concentration in effluent increased slightly (7.4 % of the initial value), indicating that the sorption capacity of 2.5 g MAS for PO_4^{3-} was beginning to be exceeded. The experiments were carried out for defining more exactly the break-through point of the cartridge (2.5 g MAS) for PO_4^{3-} and AsO_4^{3-} after passing through a large volume of water contained PO_4^{3-}, AsO_4^{3-} and AsO_3^{3-} (Figure 5.3). Phosphate concentration decreased from the initial concentration of 2500 µg/L to 33 µg/L in the first effluent 100 mL, whereas all of AsO_4^{3-} was adsorbed but AsO_3^{3-} released. Phosphate concentration in the effluent began to rise after filtering around 400 mL, but AsO_3^{3-} concentration stayed practically constant. Therefore, the break through point of one cartridge (2.5 g MAS) for quantitatively separating As(III) from As(V) and P(V) in 400 mL of solution contained 125 µg/L As_{Tot} and 2.5 mg/L P(V).

Table 5.5: Concentration of P(V) and As(III) in effleunt. 1 L of water contained PO_4^{3-} and AsO_4^{3-} passed through a cartridge with 2.5 MAS, pH: 7.0, flow rate: 50 mL/min.

Samples	P(V) (µg/L)	As(III) (µg/L)
Initial concentrations	517	134
In the first 100 ml of effluent	8.9	134
In the last 100 ml of effluent	38.3	148

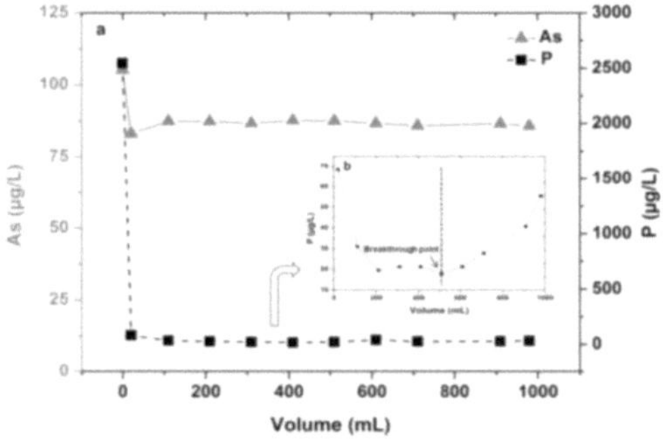

Figure 5.3: The adsorption of AsO_4^{3-} and PO_4^{3-} in one cartridge (2.5 g MAS). Passing through 1000 mL of solution containing 115 µg/L As(V) and 2500 µg/L P, where As(III)/As(V): ~1:4, pH: 7.0, flow rate: 50 mL/min.

One experiment was carried out for further improving the removal of P(V) from As(III) by MAS. After passing 500 mL of water with an initial As_{Tot}/P ratio of ~1/11 through the first cartridge contained 2.5 MAS, the total As concentration decreased slightly due to the adsorption of As(V), and the P/As_{tot} ratio changed dramatically (Figure 5.4). Passing a new the effluent, phosphate was almost completely removed (only 0.8 % of initial PO_4^{3-} remained in solution), without a significant drop in the As-total concentration, indicating an efficient retention of PO_4^{3-} on the cartridge without loss of As(III) (Figure 5.4). All the results show that MAS can completely remove AsO_4^{3-} and PO_4^{3-} from As(III), which can be applied as the first step for separation of arsenic oxyanions from water.

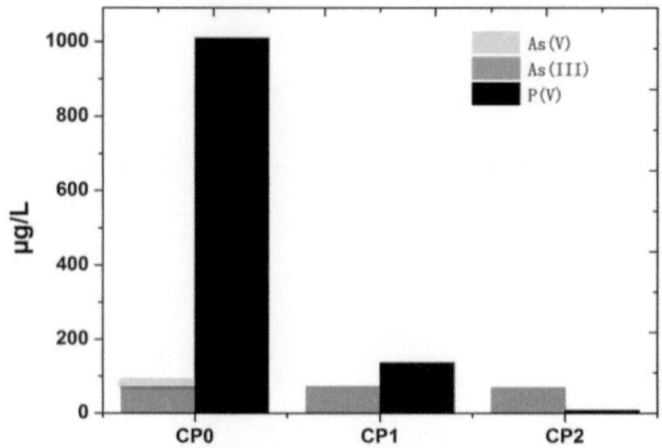

Figure 5.4: Removal of AsO_4^{3-} and PO_4^{3-} from AsO_3^{3-}. Experiment conditions: 500 mL of solution with an initial AsTot/P ~ 1/10, As(V)/As(III) ~ ¼, pH: 7.0. CP0: initial solution; CP1: first effluent; CP2: second effluent.

5.3.2.2 The desorption of arsenate from MAS adsorbed arsenate and phosphate

The experiments for desorption of AsO_4^{3-} from MAS with adsorbed AsO_4^{3-} and PO_4^{3-} are presented in Table 5.6. In addition to the adsorption of AsO_4^{3-} after shaking 30 min at 150 rpm, up to 99 % of PO_4^{3-} was also adsorbed on MAS. Sulfate concentration was around 16 mg/L in solution after adsorption process. This indicates that MAS material can release sulfur during the adsorption process of AsO_4^{3-} and PO_4^{3-} (Meng et al., 2001a). Considering the element compositions in MAS (Table 4.1), the releasing of sulfur increased assumable due to exchange with As(V) and PO_4^{3-} anions adsorbed on the filling. 92 % of As(V) was desorbed from MAS in 1 M NaOH when initial concentration of As(V) was 0.5 mg/L, and little lower desorption of As(V) when initial concentration was higher. Concurrently, adsorbed PO_4^{3-} was also released from MAS and high amount of SO_4^{2-} was detected as well in the NaOH solution, but no other common anions, like Cl^- or NO_3^-. The possible reason may be that sulfur compounds on MAS can be partly exchanged with OH^-, AsO_3^{3-} and PO_4^{3-} to release in NaOH solution.

In brief, those results show that AsO_4^{3-} and PO_4^{3-} can be separated from AsO_3^{3-}, Cl^-, NO_3^- during adsorption process. Afterwards, both of AsO_4^{3-} and PO_4^{3-} can be leached from MAS by desorbing in 1M NaOH, meantime, much

SO_4^{2-} co-occurred with them. Therefore, the tests for separation of AsO_4^{3-} from PO_4^{3-} and SO_4^{2-} were worked out after being quantitatively desorbed from MAS, which are presented in the following sections.

Table 5.6: Adsorption and desorption of As(V) and PO_4^{3-} by MAS. Adsorption: 0.5 hour, 150 rpm; desorption: 1 hour, 150 rpm, 50 mL of 1 M NaOH desorption.

As(V) (mg/L)	**P(V) (mg/L)**	As(V) (%)		PO_4^{3-} (%)		SO_4^{2-} (mg/L)	
Initial	**Initial**	Adsorbed	Desorbed	Adsorbed	Desorbed	Released after adsorption	Released after desorption
0.5	**2**	98.0	91.8	100	54.3	16.3	255
1	**2**	97.9	82.1	100	81.7	16.7	256
2	**2**	94.8	82.3	100	108	17.6	279

5.3.3 Separation of arsenate from phosphate by V-NMDG and NFO

The tests for separation of AsO_4^{3-} from PO_4^{3-} were carried out with different adsorbents, like V-NMDG, NFO and IRAN. The tested results of V-NMDG and NFO are presented in this section and the results of IRAN, which can separate AsO_4^{3-} from PO_4^{3-} successfully, are detailed showing in section5.3.4.

5.3.3.1 Separation tests by V-NMDG

V-NMDG was reported as a selective resin for adsorption of AsO_4^{3-} (Dambis et al., 2004), however, the fate of other anions, like SO_4^{2-}, PO_4^{3-} or Cl^-, was not clear. For the propose of this study, the concentrations of those anions were clarified during the adsorption and desorption of AsO_4^{3-} on V-NMDG. The results are shown in Table 5.7 and Table 5.8. In the solution with 1 mg/L of AsO_4^{3-} and presence of different concentration of PO_4^{3-}, the adsorption of AsO_4^{3-} was always high, more than 92 %, but slightly decreased as the increase of concentration of coexisting PO_4^{3-}, from 98.4 % to 92.6 % when the initial concentration of PO_4^{3-} increased from 0 to 20 mg/L. The worse phenomenon was that PO_4^{3-} was also adsorbed with AsO_4^{3-}, and released from V-NMDG together with AsO_4^{3-} during desorption process. The same results were also got when coexistence of SO_4^{2-} or Cl^- with AsO_4^{3-} (Table 5.8), more or less of those two anions were adsorbed or released together with AsO_4^{3-} during the reaction processes. Meantime, high amount of NO_3^-, which was used for protonating V-NMDG resin, released from V-NMDG. Although there is a high adsorption capacity of V-NMDG for AsO_4^{3-}, up to 80.4 mg/L, all result show that V-NMDG can not separate AsO_4^{3-} from PO_4^{3-}, SO_4^{2-} and Cl^-. Therefore, V-NMDG can not be applied to preparative separation of AsO_4^{3-} from water.

Table 5.7: Adsorption of AsO_4^{3-} and PO_4^{3-} by V-NMDG in the presence of AsO_4^{3-}. 25 mg of V-NMDG resin contacting 35 mL of 1000 µg/L As(V) solution with increasing amounts of PO_4^{3-} for 24 h. anions were desorbed from V-NMDG by 4 M HNO_3 for 24 h.

Initial PO_4^{3-} (mg/L)	**Adsorbed (%)**		**Desorbed (%)**	
	AsO_4^{3-}	PO_4^{3-}	AsO_4^{3-}	PO_4^{3-}
0	98.4	-	60.1	-
1	95.2	28.7	59.9	59.2
5	94.3	48.0	62.9	62.1
20	92.6	59.2	63.8	57.5

5.3.3.2 Separation test by using NFO

As a new adsorbent, NFO was also tested for separating AsO_4^{3-} from PO_4^{3-}. Both of AsO_4^{3-} and PO_4^{3-} were adsorbed on NFO, and the results are presented in Figure 5.5. 90 % of AsO_4^{3-} and 64 % of PO_4^{3-} were adsorbed on NFO, whereas, 87 % of AsO_4^{3-} and 55 % of PO_4^{3-} were desorbed from NFO in 1 M NaOH. As a material consisting of Fe_2O_3, there was no chance to use acid to desorb ions from the surface of NFO. Therefore, it is not possible to use NFO for preparative separation of AsO_4^{3-} from PO_4^{3-}.

Table 5.8: Arsenate Sorption by V-NMDG Resin in the Presence of PO_4^{3-} Ions. Four batch with 25 mg NMDG-V reacted in 250 mL of solution with 1 mg/L As(V), which coexisted with other anions, like 200 mg/L SO_4^{2-}, 10 mg/L of P(V) or 100 mg/L of Cl^-, shaken at 200 rpm for 24 hours, pH: 5.5-6.8.

Species present in solution	Adsorbed (%)					Desorbed (%)			
	As(V)	Cl^-	PO_4^{3-}	SO_4^{2-}	NO_3^- (mg/L)	As(V)	Cl^-	PO_4^{3-}	SO_4^{2-}
As(V) + SO_4^{2-}	22.9	-	-	32.2	95.6	90.6	-	-	2.80
As(V) + PO_4^{3-}	100	-	36.6	-	44.7	75.2	-	6.80	-
AS(V) + Cl^-	86.9	17.6	-	-	65.8	87.7	87.9	-	-
As(V) + SO_4^{2-} + PO_4^{3-} + Cl^-	26.6	0.55	5.80	30.9	94.1	94.8	-	17.5	2.10

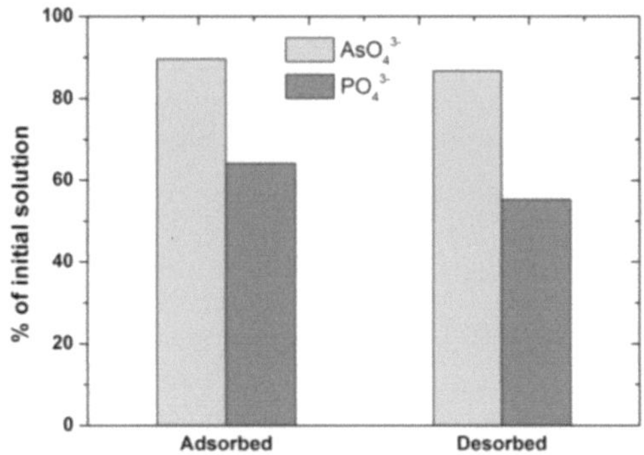

Figure 5.5: Test for the separation of AsO_4^{3-} and PO_4^{3-} by NFO. Adsorption conditions: 50 mL of solution with 1.5 mg/L As(V) and 30 mg/L P(V) was shaken with 100 mg NFO at 150 rpm for 1 hour, pH: 6.6; Desorption condition: NFO with adsorbed AsO_4^{3-} and PO_4^{3-} was shaken in 1 M NaOH at 150 rpm for 1 hour. T=21 ±1 °C.

5.3.4 Separation of arsenate from phosphate by IRAN

The aim of this section is to verify the suitability of IRAN for separation of AsO_4^{3-} and PO_4^{3-} and to develop a procedure to obtain AsO_4^{3-} in a form adequate for isotopic work. The basic idea is to load an adequate amount of AsO_4^{3-} onto MAS by passing a sufficiently large volume of groundwater through the cartridge and after leaching the anions from the column before separating AsO_4^{3-} and PO_4^{3-} by means of IRAN. This section reports the optimal conditions by which the two anions can be separated by selective stripping after being previously loaded onto IRAN.

5.3.4.1 The sorption capacity of IRAN for arsenate and phosphate

The time necessary for a quantitative adsorption of AsO_4^{3-} and PO_4^{3-} by 1 g of resin from a solution containing 1 mg/L As and 10 mg/L P has been shown to be slightly less than 1 hour (Figure 5.6). Up to about 20 minutes, the adsorption increases almost linear with time, to diminish afterwards and levels off close to 100% after one hour.

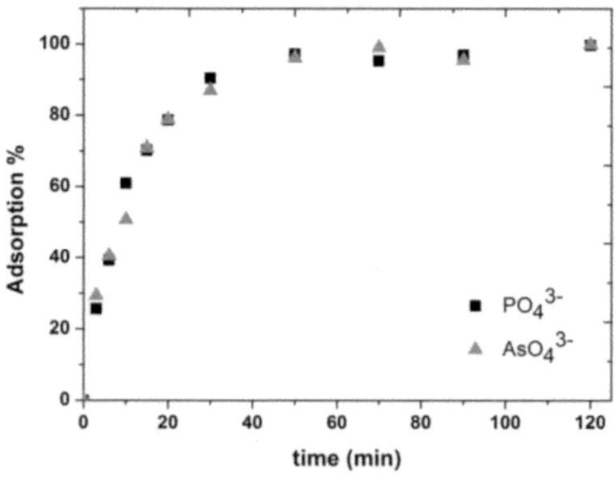

Figure 5.6: Adsorption kinetics of AsO_4^{3-} and PO_4^{3-} by IRAN. Experimental conditions: 1 g IRAN resin, 250 ml of solution with 10 mg/L P-PO_4^{3-} and 1 mg/L As- AsO_4^{3-} , pH 5.5.

Results of the experiments to find out the optimal amount of resin necessary for a complete elimination of AsO_4^{3-} and PO_4^{3-} are shown in Figure 5.7. After addition of the resin, the initial pH of the solution aliquots was around 5.5. A minimum of 0.5 g of resin is necessary for a quantitative removal of both AsO_4^{3-} and PO_4^{3-} from solutions with the specified compositions. However, considering that the actual aim of the study is to find out the optimal conditions for the quantitative extraction of AsO_4^{3-}, 0.3 g of resin is sufficient for this purpose. It also can be seen that, compared to PO_4^{3-}, the removal of AsO_4^{3-} is slightly more efficient, which is even more evident if the solution contains only arsenic. However, this may be due to its lower concentration and the competition in sorption between As and P, and not to a basically different adsorption behavior of the two anions.

The experimental data also have been used to generate the Freundlich isotherms, to define the adsorption characteristics of IRAN for of AsO_4^{3-} and PO_4^{3-} when both of them are in the solution. The observed adsorption trends for both AsO_4^{3-} and PO_4^{3-} fit well a Freundlich isotherm (1906), which is shown in Figure 5.8. While the linear correlation coefficients between the experimental data and the modeled isotherms are given in Table 5.9. The slopes of the plots of log q versus log c, for both AsO_4^{3-} and PO_4^{3-}, were very similar (Figure5.7).

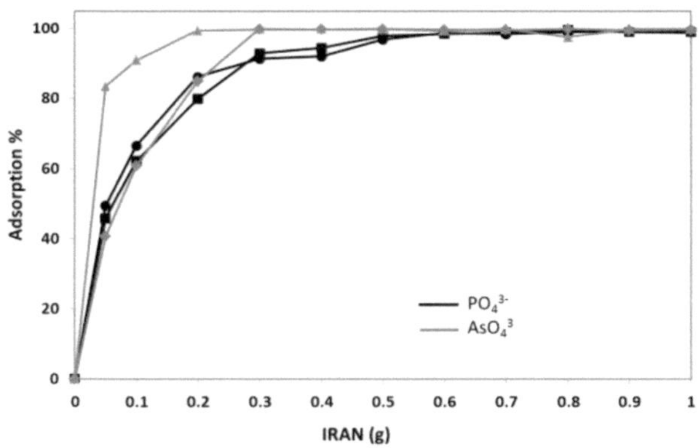

Figure 5.7: Adsorption of AsO_4^{3-} and PO_4^{3-} by different amounts of IRAN. Experimental conditions: ●: the adsorption of PO_4^{3-} in solution with 10 mg/L P; ■: the adsorption of PO_4^{3-} in solution with 10 mg/L P + 1 mg/L As; ▲: the adsorption of AsO_4^{3-} in solution with 1 mg/L As; ◆ : the adsorption of AsO_4^{3-} in solution with 10 mg/L P + 1 mg/L As. pH 5.5, stirring time 2 h at 300 rpm.

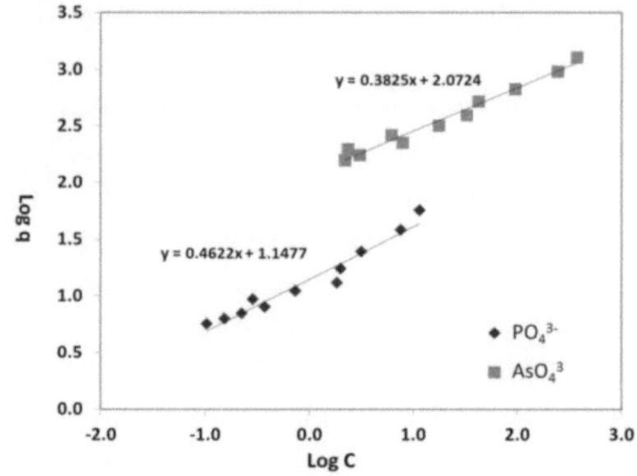

Figure 5.8: Linearized Freundlich isotherms for AsO_4^{3-} and PO_4^{3-} adsorbed by IRAN. Experimental conditions: 250 mL of solution with 1mg/L As and 10 mg/L P, pH 5.5, and different amounts of adsorbent (0.05 to 1 g).

Table 5.9: Parameters of Freundlich isotherms for the adsorption of AsO_4^{3-} and PO_4^{3-} by IRAN.

Parameters	AsO_4^{3-}	PO_4^{3-}
k	118.1	14.05
n	2.614	2.164
r^2	0.977	0.951

5.3.4.2 The effect of pH and P/As ratio on sorption

Solution batches with identical initial As(V) and P(V) concentrations (1 mg/L and 10mg/L, respectively) but different pH (the pH value from 1 to 11) were used in experiments to investigate the effect of protonation on the concomitant sorption of AsO_4^{3-} and PO_4^{3-} by IRAN (Figure 5.9). At tested pH values between 3.8 and 11, after 50 min, almost all AsO_4^{3-} and PO_4^{3-} (up to 98%) was adsorbed, with no noteworthy differences in the trend of the sorption curves at different pH. However, no adsorption was observed neither for AsO_4^{3-} nor for PO_4^{3-} at pH 1. This similarity is most likely due to the wide analogy in the protonation state and the corresponding electric charge of aqueous As(V) and P(V) species at different pH values. At a pH around 1, both As and P are uncharged as $H_3AsO_4^0$ and

$H_3PO_4^0$, respectively, to pass almost simultaneously to bi-, mono- and deprotonated oxyanions, with a corresponding change in charge as pH increases. Consequently, due to their very similar adsorption behavior, P(V)- and As(V)-oxyanions cannot be separated by a pH dependent selective sorption on IRAN.

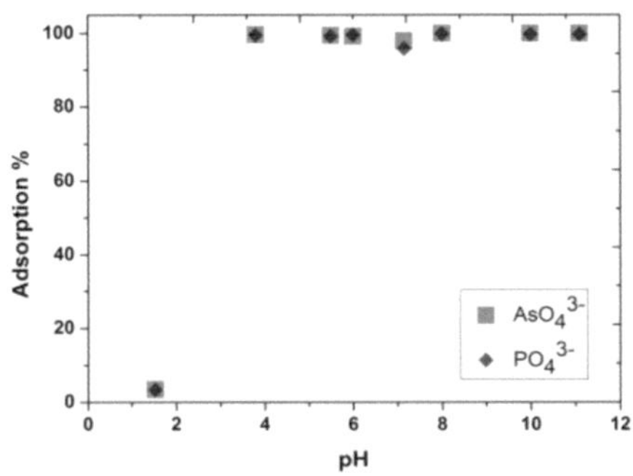

Figure 5.9: The sorption of AsO_4^{3-} and PO_4^{3-} as a function of pH. 250 mL of solution with 1mg/L As and 10 mg/L P, 1 g IRAN resin, stirred 1 hour at 300 rpm.

Table 5.10: Adsorption behavior of IRAN under different ratios of As/P.

Ratio of As/P	As (mg/L)/P (mg/L)	P (V) adsorbed (%)	As(V) adsorbed (%)
1:1	0.2/0.2	100.0	96.6
1:2	0.2/0.4	99.6	98.3
1:10	0.2/2.0	97.5	98.2
1:30	0.2/6.0	97.1	97.8

Experimental conditions: 250 mL of solution with different concentration of AsO_4^{3-} and PO_4^{3-}, 1 g IRAN resin, stirring time 1 hour at 300 rpm.

The effect of the P/As ratio on sorption was investigated by keeping the As concentrations at 0.2 mg/L, and increasing the P concentration from 0.2 to 6.0 mg/L. In a P to As ratio range of 1 to 30, no significant differences in the adsorption of the two anions was observed (Table 5.10). The sorption is close to 100 % for both, but while the adsorption of AsO_4^{3-} is constantly around 97. 7± 0.7 %, the sorption efficiency for PO_4^{3-} decreases slightly from 100 % to 97.1 %, as P concentration increases from 0.2 to 6.0 mg/L. These results indicate that the initial concentration ratios do not affect substantially the adsorption of AsO_4^{3-} and PO_4^{3-}.

5.3.4.3 Adsorption of other anions during adsorption process

Solutions containing 100 mg/L of Cl^-, SO_4^{2-} and NO_3^- were used to investigate the effect of concurrent anions on sorption. In the presence of these anions, AsO_4^{3-} and PO_4^{3-} concentrations still decreased to values close to zero (Figure 5.10), but in the same time Cl^-, SO_4^{2-} and NO_3^- were also co-adsorbed. While the retention of SO_4^{2-} and NO_3^- were practically quantitative, Cl^- concentrations in the residual solution were still relatively high. This may be due to some remnant (still high) Cl^- contents in the resin, which occurred because the commercially available IRAN used in the experiments was prepared from its chloridic form. The elimination of Cl^- from the solution especially is important, since high amounts of AgCl can be formed during the subsequent precipitation of Ag_3AsO_4 and complicates the isotopic measurements. Although coexisting anions commonly present in groundwater seem not to affect significantly the adsorption behavior of As(V)- and P(V)-oxyanions, high concentrations of other anions may potentially influence negatively the separation of AsO_4^{3-} and PO_4^{3-} during their selective stripping from the resin.

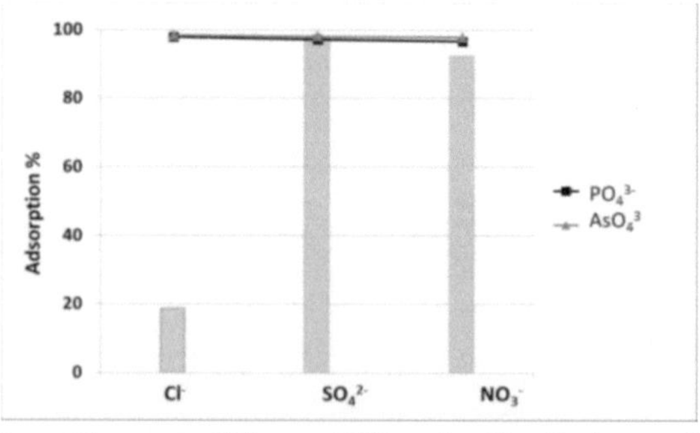

Figure 5.10: Reaction conditions: 250 mL of solution with 1 mg/L As and 10 mg/L P, 100 mg/L of Cl^-, SO_4^{2-} and NO_3^- in each batch, 1 g IRAN resin, stirred 1 hour at 300 rpm.

5.3.4.4 Selective desorption of arsenate and phosphate from IRAN

The actual separation of AsO_4^{3-} and PO_4^{3-} was achieved by their selective desorption from the resin. In these experiments, different eluents (NaOH, HCl, CH_3COOH, HNO_3) and concentrations (0.5, 1.0, 2.0 mol/L) were applied to find out the optimal conditions to separate the anions (Figure 5.11). Under alkaline conditions (250 ml of 1 mol/L NaOH), both AsO_4^{3-} and PO_4^{3-} were stripped quantitatively from the resin. Using a weak acidic solution (0.5 mol/L CH_3COOH) only 30 % of the adsorbed AsO_4^{3-} and 45 % of the adsorbed PO_4^{3-} can be recovered, but the desorption increased to 60-65 % for both anions when the acid concentration was raised to 1 mol/L. An effective separation can be achieved, however, with a strong acidic solution (1 mol/L HNO_3 and HCl, respectively), by which 99 % of the AsO_4^{3-}, but only maximum 5 % of PO_4^{3-} was desorbed from the resin (Figure 5.12).The efficiency of the separation can be further improved when the concentration of HCl or HNO_3 was increased to 2 mol/L (Figure 5.12). Under these conditions, PO_4^{3-} was below detection limit (<0.005 mg/L) in the leachate, whereas AsO_4^{3-} was released to 99 %.

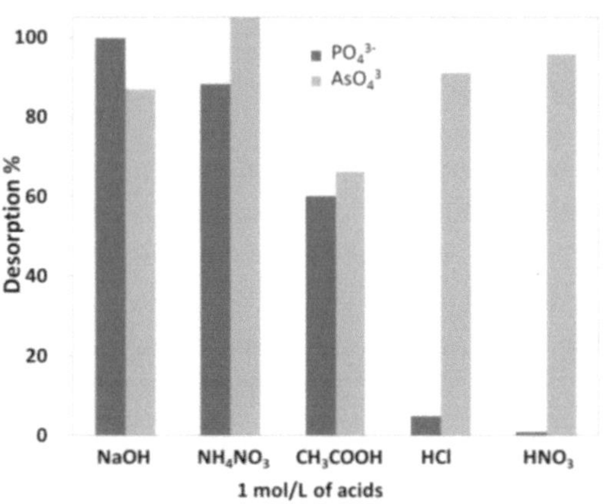

Figure 5.11: Efficiency of selective AsO_4^{3-} and PO_4^{3-} desorption from IRAN with different eluents in 1 mol/L concentrations.

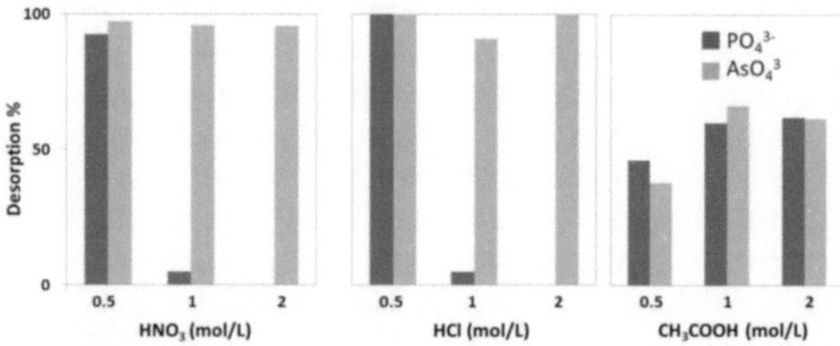

Figure 5.12: Efficiency of AsO_4^{3-} and PO_4^{3-} separation using different concentrations (mol/L) of HNO_3, HCl and CH_3COOH as eluents.

5.3.4.5 Evaluation of adsorption efficiency by scanning electron microscopy combined with X-ray spectroscopy (SEM-EDX)

To confirm the correctness of the separation process, SEM-EDX was used to record and document changes in elemental composition on the surface of the resin grains during the different steps of the separation procedure (Figure 5.13). In line with the organic matrix of IRAN (styrene-divinilbenzene gel), the unloaded grains are characterized by high peaks for carbon and oxygen (Figure 5.13a). The presence of chlorine is due most likely to leftovers from the manufacturing process, during which the Cl^- form of the resin was transformed into the (OH) form. Further on, by means of EDXRFA small amounts of Ca, K, S and Al were detected in the resin (table 4.3). After one hour of shaking with a solution with 1 mg/L of As(V) and 10 mg/L P(V), the EDX spectrum shows clearly the loading of these elements on the surface of the grains (Figure 5.13b). The leaching with 250 mL of 2 M HNO_3, causes the disappearance of arsenic and also of Cl peaks (Figure 5.13c), but not of the P peak. The retention of P in form of a distinct surface complex is indicated by the occurrence of particles with high P content adhered to the surface of the beads (Figure 5.14).

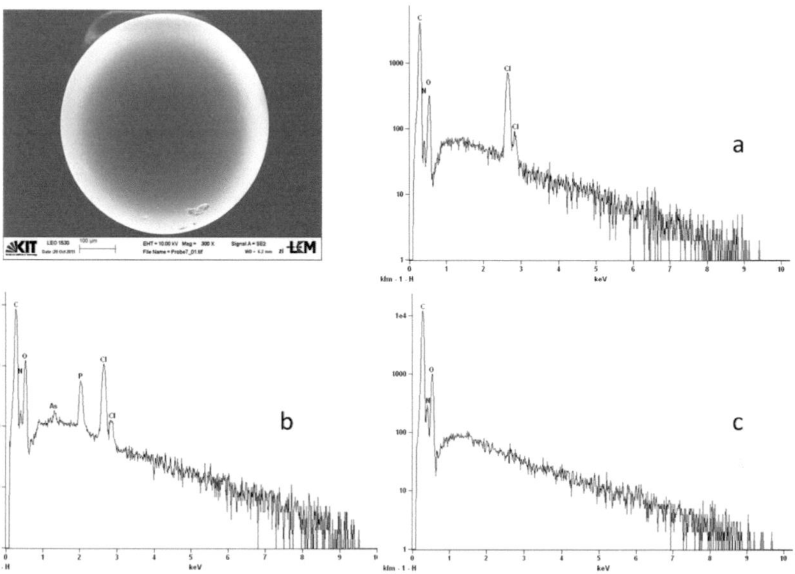

Figure 5.13: SEM micrograph of an IRA-400 bead and EDX-spectra produced by the grain surface in different steps of the separation experiment: (a) before adsorption; (b): after loaded with both AsO_4^{3-} and PO_4^{3-}; (c) after desorption with 2 M HNO_3.

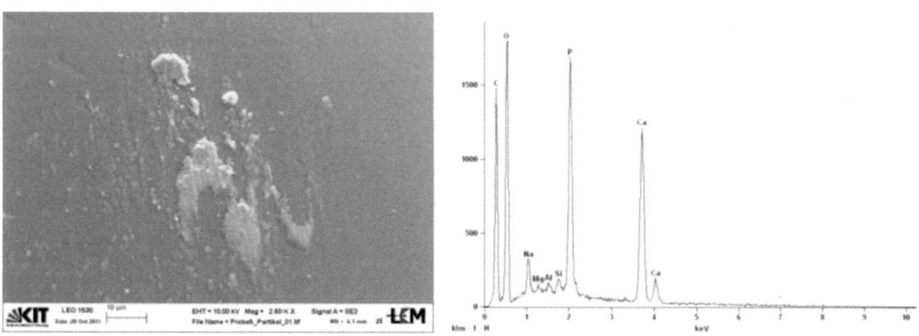

Figure 5.14: SEM micrograph of the IRA-400 bead surface and EDX spectrum of a particle adhering to the surface of the resin after eluted with 2 M HNO_3.

5.3.5 Separation of arsenate from sulfate

Considering IRAN also adsorbs SO_4^{2-} beside of AsO_4^{3-} and PO_4^{3-} (section 5.3.4.5), the experiment was carried out for checking the concentrations of SO_4^{2-} after desorption of AsO_4^{3-} in 2 M HNO_3 (Table 5.11). After quantitative adsorption of those three anions, adsorbed AsO_4^{3-} and SO_4^{2-} quantitatively was released in 2 M HNO_3, but PO_4^{3-} was still bound on IRAN and not detected in the leachate. It indicates that IRAN can be used to separate AsO_4^{3-} from PO_4^{3-}, but not from SO_4^{2-}. The further experiments were worked out for separation of AsO_4^{3-} from SO_4^{2-}.

Table 5.11: Concentration change of SO_4^{2-} during separation of AsO_4^{3-} from PO_4^{3-}. Adsorption: 100 ml of solution with AsO_4^{3-}, PO_4^{3-} and SO_4^{2-} was shaken with 0.5 g IRAN at 150 rpm for 1 hour, pH: 6.5; desorption: IRAN with adsorbed anions were shaken in 100 mL of 2 M HNO_3 at 150 rpm for 1 hour.

Samples	As(V (µg/L)	PO_4^{3-} (mg/L)	SO_4^{2-} (mg/L)
Initial concentration	1619	16.9	100
Concentration after adsorption	88.8	2.08	<5
Concentration after desorption	1510	<0.005	92.4

Three adsorbents, BYC MHZR and NFO, were tested for separation of AsO_4^{3-} from SO_4^{2-}. Changes in pH, in amount of AsO_4^{3-} and amount of coexisting SO_4^{2-} on the three adsorbent are shown in Figure 5.15. From a solution with 1300 µg/L As(V), 98.9 %, 99.7 % and 87.1 % of AsO_4^{3-} was adsorbed in 1 hour (Wasay et al., 1996), 3 hour (Suzuki et al., 2000) and 1 hour, on BYC, MHZR and NFO, respectively (Figure 5.15b). The pH value of the solution shaken with MHZR decreased during the experiment from 5.95 to 4.98, which was most possibly due to the release of H^+ from the surface of the resin to the solution. The pH increased from 5.95 to 7.79 in the experiment with BYC, with which the removal of AsO_4^{3-} by precipitation together with adsorption takes place in the pH range, due to dissolution of BYC (Wasay et al., 1996). In the experiment with NFO, pH increased from 5.95 to 6.79, possibly since $H_2AsO_4^-$ was less in the solution after adsorption than in initial one, thus the solution went back to neutral condition (Figure 5.15a). The significant phenomenon is that no change in concentrations of SO_4^{2-} in effluent has been recorded during the adsorption of AsO_4^{3-} by these three adsorbents (Figure 5.15c). Whereas, 84.0 %, 85.1 % and 79.9 % of AsO_4^{3-} was desorbed from BYC, NMHR and NFO by using 1 M, 0.5 M and 1 M NaOH, respecitively (Figure 5.15b). For the purpose of the study, this indicates that all three tested adsorbents can be used to separate AsO_4^{3-} from SO_4^{2-}.

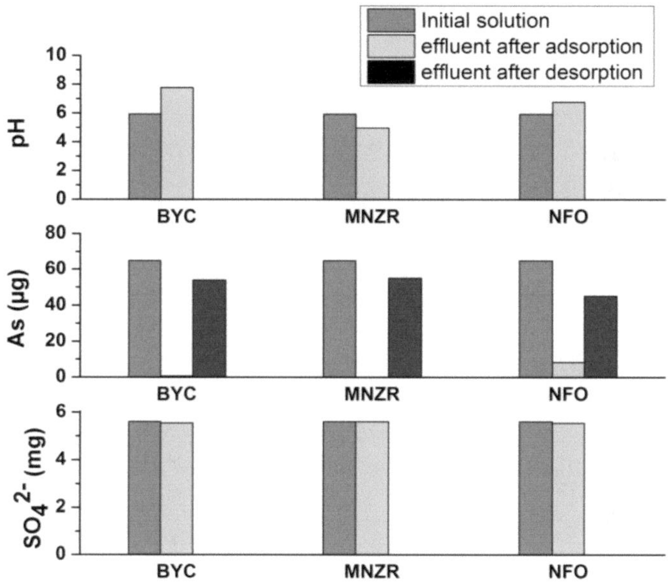

Figure 5.15: Changes in pH and concentrations of pH, AsO_4^{3-} and SO_4^{2-} during the nearly quantitative adsorption and desorption of AsO_4^{3-}on MHZR (1 g), BYC (25 mg) and NFO (10 mg). Shaking rate: 150 rpm; reaction times: MHZR: 3 hours for adsorption and 1 hour for desorption; BYC: 1 hours for both of adsorption and desorption; NFO: 1 hour for both of adsorption and desorption, respectively. pH value was not measured after desorption of AsO_4^{3-} by strong acid or alkali solution.

5.4 Summary

Experiments carried out with different adsorbents have shown that some of them can be used successfully for the preparative separation of AsO_4^{3-} from other common existed anions like PO_4^{3-}, SO_4^{2-} and Cl^-, for the mandatory steps in determination of the isotope composition of AsO_4^{3-}-oxygen. The role of each adsorbent for separation of AsO_4^{3-} can be summarized as follows and the whole procedure for separation of AsO_4^{3-} from natural water can be described as in Figure 5.16:

1) First step, MAS, whose adsorption capacity of AsO_4^{3-} was 0.562 mg/L, can be applied for adsorbing of AsO_4^{3-} and PO_4^{3-} but not anions like AsO_3^{3-}, Cl^- and SO_4^{2-} from water sample. After the adsorption , AsO_4^{3-} and PO_4^{3-} on MAS were released in a small volume of 1 mol/L NaOH solution by shaking 30 min. This allows not only the removal of disturbing anions (other oxianions, Cl^-), but also a preconcentration of the very low AsO_4^{3-} contents, as descoveried in groundwaters. However, sulfur compound, which was one existed compostion in MAS, was also dissolved in 1 M NaOH beside with those two anions.

2) Second step, after eliminating the disturbed anions like AsO_3^{3-} and Cl^- by MAS, AsO_4^{3-} and PO_4^{3-} extracted from large volumes of water can be loaded subsequently onto 0.3-0.5 g of IRAN, whose adsorption capacity of AsO_4^{3-} is 43.1 mg/L. After adjusting pH of solution to >3.8, AsO_4^{3-} can be loaded (together with PO_4^{3-} and SO_4^{2-}) on IRAN beads, from which it can be selectively stripped with 2 mol/L HNO_3. In this step SO_4^{2-} is stripped, too. While PO_4^{3-} concentrations in the leachate were detected to be below detection limit, AsO_4^{3-} and SO_4^{2-} passed quantitatively into solution. In this step, AsO_4^{3-} and SO_4^{2-} were

quantitatively separated from PO_4^{3-}. In order to avoid the release and co-precipitation of disturbing anions, like Cl^-, the resin must be thoroughly clean-washed by diluted concentration (like 0.25 mol/L) of NaOH before use.

3) Third step, after leaching from IRAN, three adsorbent, BYC, MHZR and NFO can be applied for separation of AsO_4^{3-} from SO_4^{2-}. Under the optimal conditions, the adsorption reaction with BYC, NMHR and NFO was 1 hour, 3 hour and 1 hour, respectively, and AsO_4^{3-} can be quantitatively adsorbed on BYC and NMHR, and 87.1 % adsorbed on NFO. After the adsorption process with those three adsorbents, the concentration of SO_4^{2-} in the leachate was almost the same as it was in the initial solution. In the desorption process, 84.0 %, 85.1 % and 79.9 % of the adsorbed AsO_4^{3-} on BYC, NMHR and NFO were stripped in1 M, 0.5 M and 1 M NaOH, respectively. Although the role of NFO for separation of AsO_4^{3-} from SO_4^{2-} was less effective than the other two, NFO still can be an effective adsorbent to be used in this separation step.

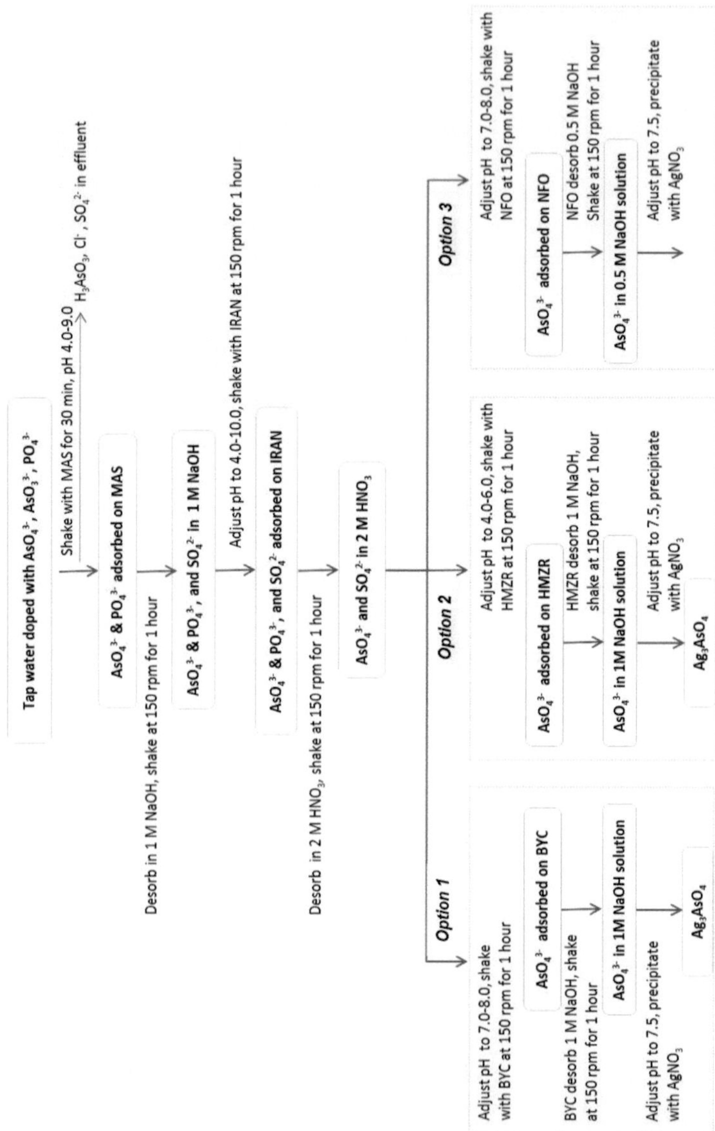

Figure 5.16: Schematic process-flow with three options for extraction of AsO_4^{3-} from arsenic contaminated natural water samples. (working temperature: 21±1°C).

6 Separation of arsenite from water

6.1 Introduction

Arsenite is the dominant specie in groundwater and the determination of $\delta^{18}O$ vaule in AsO_3^{3-} is especially important in the potential field applications. The study of the isotopic composition of oxygen in AsO_3^{3-} may be more potentially highly instructive in understanding the mechanisms of arsenic mobilization in groundwater. Like the scheme for the preparative separation and extraction of AsO_4^{3-} from natural water (chapter 5), the focus of this chapter is to develop an efficient method for separation of AsO_3^{3-} from all common oxygen bearing anions occurring in natural water. Because the presence of Cl in solid phase together with AsO_3^{3-} does not affect the measurement of oxygen isotope in silver arsenate, the first test for extraction of AsO_3^{3-} from water did not take the removal of Cl^- into concern.

However, due to too much AgCl can dilute Ag_3AsO_3 concentration in the precipitates after extraction procedure, the successful method for extraction of AsO_3^{3-} for measurement of $\delta^{18}O$ vaule must remove Cl- from AsO_3^{3-}. The approach adopted in this study starts also from the separation of AsO_3^{3-} from AsO_4^{3-} and PO_4^{3-} by alumino-silicate based sorbent (MAS) (Meng & Wang, 1998). While AsO_3^{3-}, NO_3^- and Cl^- pass more or less unhindered through and AsO_4^{3-} and PO_4^{3-} concentrations are below detection limit in the filtrate and SO_4^{2-} even increased (Table 4.6). As shown before (chapter 4), HCO_3^- can be easily removed as CO_2 by lowering the pH of the filtrate below 4.3, and the presence of NO_3^- is not critical. A challenging task here is, however, the quantitative elimination of SO_4^{2-} and Cl^- anions from the filtrate, without changing the oxidation state of AsO_3^{3-}. If present, these anions impede the pre-concentration and extraction of pure AsO_3^{3-}, as required by the oxygen isotopic work.

At neutral pH, the thermodynamically stable form of As(III) is the undissociated arsenious acid (H_3AsO_3) (Figure 6.1). The pK_{a1} of H_3AsO_3 is 9.1 and hence the mono-anionic $H_2AsO_3^-$ is likely to be responsible for the adsorption of As(III) in alkaline range (Cotton & Wilkinson, 1962). Many techniques used for the removal of arsenic, prior to adsorption, are oxidizing As(III) into As(V), since arsenic acid is dissociated over a far larger pH range than arsenous acid (Mohan & Pittman, 2007). However, such techniques are inappropriate for our purpose, because they would evidently alter the isotopic composition of oxygen. A further requirement is the enrichment of the relatively low initial AsO_3^{3-} concentrations during the separation to gain enough mass for oxygen isotope analysis. In most convenient way, this can be done by loading the arsenic content of large volumes of water on a sorbent, from which AsO_3^{3-} can be stripped within a small volume of eluent. Consequently, a material suitable for this purpose must have a high adsorption capacity for AsO_3^{3-}, but not for SO_4^{2-} and Cl^-.

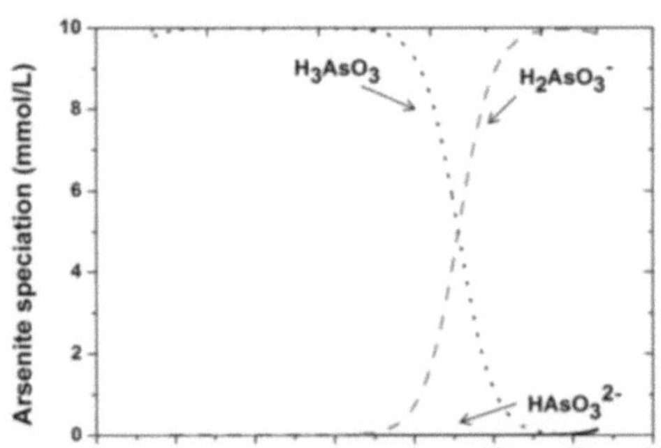

Figure 6.1: Speciation of As(III) as a function of pH for a total As(III) concentration of 10 mmol/L (by PHREEQC program with water4f.dat).

Taking into account the above constraints, the beginning test was to develop a separation method to remove SO_4^{2-} and HCO_3^-. In this first step the removal of Cl^- was not taken into consideration to reduce the complexity of this task. The results are shown in section 6.3.1. However, the results from this test show that the presence of much Cl^- significantly affected the further work of oxygen isotope. Therefore, the tests had to be expanded and were then focused on the removal of SO_4^{2-}, HCO_3^- and Cl^- simultaneously. Several adsorbents, which have been selected before, BYC, MHZR, NFO and IRAN were tested for this proposal.

6.2 Experiments

6.2.1 Precipitation of As(III) direct from natural water.

Considering the oxygen value in AsO_3^{3-} is not affected by coexisting Cl^-, the goal of this experiment was to find the feasibility for the precipitation of AsO_3^{3-} beside Cl^- after removing HCO_3^- and SO_4^{2-}. The tested solution was prepared in 1 L of mineral water with additional of 1 mg $NaAsO_2$. The mineral water contained also 60.7 mg/L Cl^-, 194 mg/L SO_4^{2-} and 337 mg/L HCO_3^- and pH value was 7.73 (Table 6.1). The removal of HCO_3^- was first done, by addition of 390 µL of 65 vol% HNO_3, which adjusted the pH value of the solution to 4.2 to transform all HCO_3^- into CO_2. Then, SO_4^{2-} was removed by precipitating with 30 ml of 16 mg/L $Ba(NO_3)_2$ as $BaSO_4$, which is the widely known method to quantitatively precipitate SO_4^{2-} from water. The reaction process was taken for 5 min and $BaSO_4$ was collected on weighted filter (0.45 µm) and then washed with 21 mL Mill-Q water for several times. After filtering $BaSO_4$, All the solution was collected to avoid the loss of AsO_3^{3-}. 10 mL of 33.0 g/L $AgNO_3$ was added to the filtrate for precipitating AsO_3^{3-} as Ag_3AsO_3, meantime, AgCl precipitated as well. The whole reaction took 10 min and the pH of the solution was adjusted to 7.51 by 200 µl of 28 % NH_4OH. The precipitates ($AgCl+Ag_3AsO_4$) were filtered with a weighted filter (0.45 µm). All precipitates were dried at 50 °C over night.

Table 6.1: Composition of the mineral water (Schönborn).

Ions	**K^+**	**Na^+**	**Mg^{2+}**	**Ca^{2+}**	**Cl^-**	**SO_4^{2-}**	**HCO_3^-**
mg/L	11.2	1.7	21.7	187	60.7	194	337

6.2.2 Experiment for adsorption isotherm

The adsorption isotherms were obtained by the batch method employing a constant weight of different adsorbents and 50 mL of the As(III) solution at different concentrations (0.01 mg/L, 0.05 mg/L, 0.1 mg/L, 1 mg/L, 5 mg/L, 10 mg/L, 50 mg/L, 100 mg/L, 250 mg/L, 500 mg/L and 1500 mg/L) and fixed amounts of adsorbents in 50 mL bottles for 24 hours at 150 rpm at room temperature (21±1°C). Table 6.2 presents the optimal experimental conditions for the adsorption capacity of the adsorbents, which were mainly applied for the separation of As(III) from Cl^- and SO_4^{2-}. Those solutions containing As(III) were buffered at a pH optimum for adsorption by HNO_3 and NH_4OH.

Table 6.2: Reaction conditions of adsorption isotherm experiments for AsO_3^{3-} by different adsorbents. Volume: 50 ml; T: 21±1°C

Anion	**Adsorbents**	**Weight (g)**	**pH (experimental value)**	**pH (optimum range)**	**Reaction time (hour)**	**Shaking rate (rpm)**
	BYC	0.025	9.96	9.5-10.5	24	150
As(III)	MHZR	0.1	9.96	9.5-10.5	24	150
	NFO	0.01	9.96	7.0-9.0	24	150

6.2.3 Selective separation experiments

The adsorption experiments were done in 1 L of solution with 1 mg/L As(III), 100 mg/L of Cl^- and 100 mg/L SO_4^{2-} by shaking at a rate of 150 rpm. The mixed solution was with different adsorbents under different pH values, reaction time and other parameters, aiming to define the optimal conditions for adsorption of As(III) from other co-occurring oxyanions. The optimal conditions for the adsorption of As(III) on BYC and MHZR have been more or less clarified by Wasay et al. (1996) and Suzuki et al. (2000), and the adsorption properties of NFO for As(III) are described in this study. As(III) was desorbed from materials by using different desorption agents, such as different concentrations of HNO_3 and NaOH. All experiments were performed at room temperature 21±1°C. The detailed experimental conditions are specified in the results part and in the caption of the figures or tables. Each experiment was repeated at least three times to prove and improve the precise and correction of the experimental results.

6.3 Results and discussion

6.3.1 Separation and preconcentration of arsenite from artificial groundwater

6.3.1.1 Effect of Ba^{2+} on arsenite precipitation

The first step of the test was done under alkali condition (pH~10), which was tested that dissociated AsO_3^{3-} was not removed from the solution during the precipitation of SO_4^{2-} with $Ba(NO_3)_2$ (Figure 6.2). When $Ba(NO_3)_2$ was added to a solution containing only AsO_3^{3-}, more than 94% of AsO_3^{3-} was still in solution. However, when AsO_3^{3-} and SO_4^{2-} co-existed in water, SO_4^{2-} was quantitatively removed as $BaSO_4$, but only 63% of AsO_3^{3-}, remained in solution. The results reveal that most As(III) can coexist with with Ba^{2+} in solution when only pure As(III) existed in solution, but around 40 % of As(III) can be precipitated together SO_4^{2-} with Ba^{2+} in the solution with both As(III) and SO_4^{2-}.

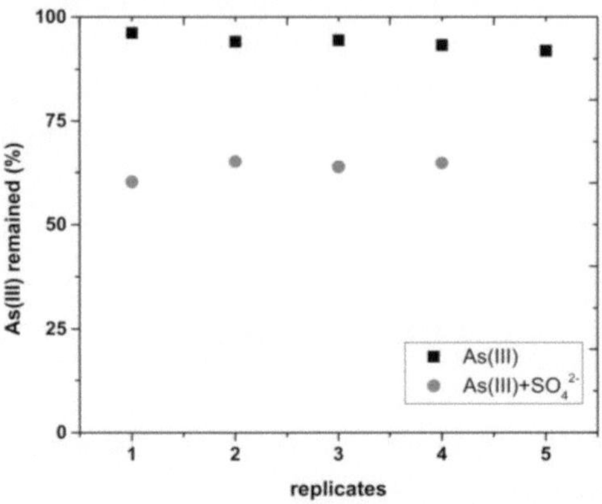

Figure 6.2: Fraction of dissolved asenite after adding $Ba(NO_3)_2$ in a solution with and without sulfate. Experimental conditions: 50 mL of two batch solution (1) 60 mg/l pure As(III) and (2) 30mg/L As(III) + 1200 mg/L SO_4^{2-}; reacted with $Ba(NO_3)_2$: 5 min. each batch was repeated for several times, pH:10, T=21 ±1 °C.

6.3.1.2 Precipitation of arsenite from natural mineral water

Although the results above show that a part of As(III) was precipitated with Ba^{2+} when SO_4^{2-} was presence in solution, it still makes sense to test whether the method can apply to separation low concentration of AsO_3^{3-} from water. The whole procedure for separation of As(III) from other oxyanions like HCO_3^- and SO_4^{2-} from natural water was still tested following the steps in Figure 6.3. The results of each step of the experiments are described in Figure 6.4. After reducing pH of solution below 4.3 and reacting for 5 min, the concentration of HCO_3^- was below the detection limit. At the low pH (<4.3), SO_4^{2-} was then totally removed by additional of excess of $Ba(NO_3)_2$. During this reaction, the significant phenomenon was that there no As(III) precipitated with Ba^{2+}. It was possible because As (III) existed as H_3AsO_3, which is more soluble in the solution with high H^+ concentration, was prevented to precipitate with Ba^{2+} by the presence of Cl^-. After adding $AgNO_3$ in excess in the effluent solution, the concentrations of both As(III) and Cl^- were under the detection limit in the solution after the precipitate was filtered. Those results show that HCO_3^- and SO_4^{2-} were quantitatively removed, while AsO_3^{3-} was recovered to 99% as Ag_3AsO_3 together with AgCl in the end of experiments (Figure 6.4). The purity of the precipitate from the two steps was checked by EDXRF (Table 6.3), which confirmed that the first precipitate ($BaSO_4$) contents large amount of Ba, S and little amount of Cl, whereas the second precipitate (Ag_3AsO_3 + AgCl) is composed entirely of Ag_3AsO_3 and AgCl, and the amount of AgCl was around 520 times more than the amount of Ag_3AsO_3 in the mixed precipitate when the concentration of Cl^- and As(III) in the original mineral water was 60.7 mg/L and 1 mg/L As(III), respectively. Although the presence of AgCl in the precipitate with Ag_3AsO_3 did not affect the value of oxygen isotope in AsO_3^{3-}, the coexistence of the high amount of AgCl in the mixed precipitates can lead much more weight of precipitates are required for the measurement of oxygen isotope. Therefore, the procedure in further separation experiments must reduce the concentration of Cl^- in order to avoid the formation of AgCl in the precipitate, which can dilute the amount of silver arsenate in the precipitate.

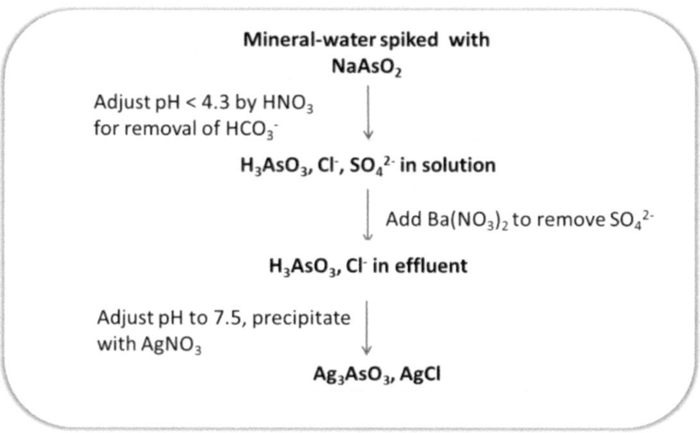

Figure 6.3: The flowchart for oxyanions (HCO_3^- and SO_4^{2-}) removal from As(III) in natural water.

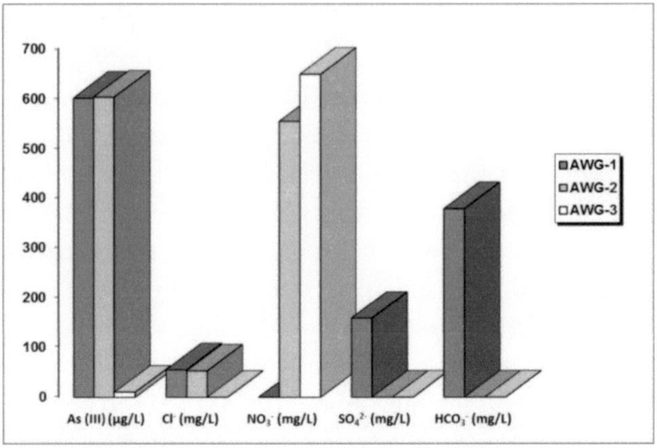

Figure 6.4: The concentration of AsO_3^{3-} and other anions in solution after separation following the scheme in Figure 6.3. • AWG-1: 1 L of mineral water spiked with 1 mg/L of As(III); • AWG-2: solution after adjusting pH to < 4.3, and reaction with $Ba(NO_3)_2$ in excess; • AWG-3: solution after precipitation with $AgNO_3$.

Table 6.3: The element content in precipitation after filtering from AWG-2 and AWG-3. AWG-2: solution after adjusting pH to <4.3, and reaction with $Ba(NO_3)_2$ in excess; • AWG-3: solution after precipitation with $AgNO_3$.

Sample identification	AWG-2 ($BaSO_4$)	AGW-3 ($AgCl+Ag_3AsO_3$)
S (ppm)	42100	-
Cl (ppm)	402	60500
As (ppm)	-	117
Ag (ppm)	-	19500
Ba (ppm)	10900	312

6.3.2 Separation of arsenite from chloride by IRAN

IRAN, which have been used for the successful separation of As(V) from PO_4^{3-}, can also adsorbmore than 85 % of As(III) (Table 4.8). It can adsorb As(V), PO_4^{3-}, Cl^-, NO_3^- and SO_4^{2-} at pH above 3.8, and As(V), PO_4^{3-}, Cl^- and SO_4^{2-} can be desorbed from IRAN under neutral and alkali conditions (Mustafa et al., 1994; 2004), whereas As(V) and SO_4^{2-} were together desorb in 2 M HNO_3 (Figure 5.10)., However, The separation of As(III) from Cl^- by IRAN during desorption process was not clear. Therefore, a test was carried out here for separation of AsO_3^{3-} from Cl^-. The results are shown in Table 6.4. More than 85.7 % of As(III) with 68.0 % of Cl^- were adsorbed on IRAN at the pH value of 7.1 and 11.1, and both of As(III) and Cl^- were desorbed in 2 M HNO_3, which indicate that IRAN can not be used for separation of As(III) from Cl^-.

Table 6.4: the data for the separation of AsO_3^{3-} from Cl^- by IRAN. Experimental conditions: adsorption: 50 mL of solution with 500 ug/L As(III) and 100 mg/L Cl^- was shaken with 0.5 g IRAN at 150 rpm for 1 hour; desorption: IRAN with adsorbed anions was shaken in 20 mL of 2 M HNO_3 for 1 hour.

pH	Adsorbed As(III) (%)	Desorbed As(III) (%)	Adsorbed Cl^- (%)	Desorbed Cl^- (%)
7.1	85.7	99.1	68.0	99.2
11.1	89.3	96.7	70.0	98.3

6.3.3 Separation of arsenite from chloride and sulfate

Three materials were considered closer in experiments for the separation of AsO_3^{3-} from the filtrate, which passed the cartridge of MAS (Meng and Wang, 1998), to reduce the concentration of Cl^- and SO_4^{2-} during the last precipitation step: (i) basic yttrium carbonate (BYC), (ii) porous polymer beads loaded with monoclinic hydrous zirconium oxide (MHZR) and (iii) nanoparticulate Fe_2O_3 (NFO). BYC has been reported to remove both As(III) and As(V) as well as PO_4^{3-} under laboratory conditions (Wasay et al, 1996; Haran et al, 1997), while MHZR was developed as an effective adsorbent for AsO_4^{3-}, AsO_3^{3-}, PO_4^{3-}, F^-, and Pb(II) by *Suzuki et al.* (1997; 2000). Both of these materials have been shown to adsorb As(III), but there is no information about the fate of the other anions relevant in this study during the sorption. The efficiency of arsenic removal via ferrihydrite co-precipitation was verified for and proved under different Fe/As ratios, pH values and with different structural characteristics, but again, the selectivity of these materials with respect to Cl^- and SO_4^{2-} is not known (Richmond et al, 2004; Tang et al, 2011). The iron oxide (NFO) used in this study is a nanoparticulate Fe_2O_3 produced by flame synthesis in our laboratory, whose properties for As(III) adsorption is presented here for the first time.

6.3.3.1 Effect of pH on adsorption

According to *Wasay et al.* (1996), the optimal adsorption of As(III) on BYC occurs in a pH range between 9.8-10.5, whereas on MHZR between 9.0-10.0 (Suzuki et al., 2000). In this pH range, the main As(III) species in solution, which may participate in the exchange reactions, is the mono-valent $H_2AsO_3^-$ anion (85-100 %) (Figure 6.1). In BYC, during the exchange, $H_2AsO_3^-$ substitutes CO_3^{2-} as $Y(OH)(H_2AsO_3)_2$, leading to the release of CO_3^{2-}, whereas on the MHZR, the $H_2AsO_3^-$ is bond as $ZrOH_2^+(H_2AsO_3)^-$ after the ion exchange occurs.

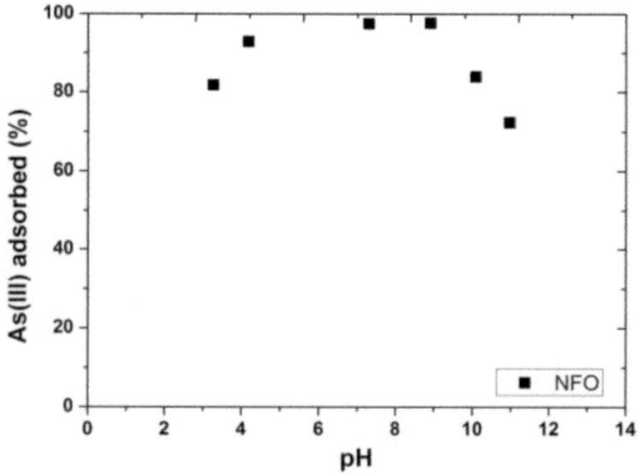

Figure 6.5: Sorption of As(III) on NFO as a function of pH. 50 mL solution with 1 mg/L of As(III) shaken at 150 rpm with 25 mg of NFO for 24 hours. The pH was periodically adjusted with 65 vol% HNO_3 or 32 vol% NaOH.

Since NFO is a new material in this study, the effect of pH on AsO_3^{3-} adsorption by NFO was investigated in a range between 3 and 11 (Figure 6.5). Maximum adsorption occurred between pH 7 and 9. Although the adsorption of arsenic on the NFO occurs in a relatively large pH range (Figure 6.5), the Fe_2O_3 started to dissolve at pH below 5 or to transform into

FeOOH under stronger alkaline conditions (pH > 9) (Lee et al., 2007). These processes may lead to a change in the specific surface area of the material and/or habitus of the nanoparticles, causing a decrease in the adsorption rate and adsorption capacity of the material. Since in a pH range between 7.0 and 9.0 As(III) occurs as uncharged $H_3AsO_3^0$, As(III) adsorption on NFO is a physisorption, instead of ion-exchange.

6.3.3.2 Adsorption isotherms

In order to estimate the adsorption constants, adsorption densities and adsorption maxima for the tested materials, the experimental adsorption data were fitted by the isotherm models of Langmuir (1961) and Freundlich (1906) (Figure 6.6).

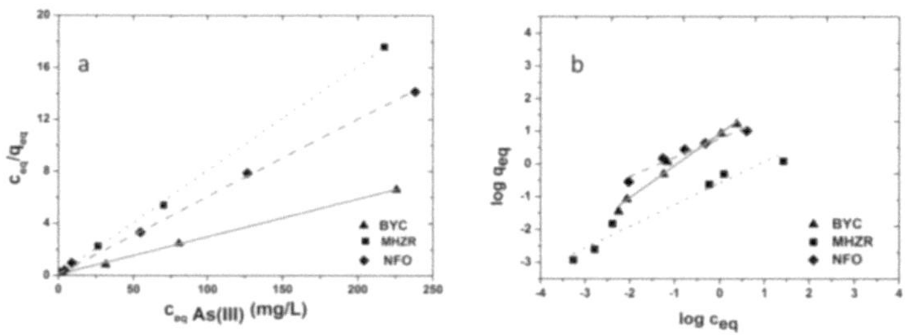

Figure 6.6: As(III) adsorption isotherms on BYC, MHZR and NFO. a: Langmuir isotherm; b: Freundlich isotherm. Experimental conditions: 50 mL batches, with As(III) concentrations in the range between 0.01 mg/L and 250 mg/L. Shaking rate: 150 rpm; duration: 24 hours. BYC: 25 mg, pH: 9.8-10.2; MHZR: 100 mg, pH: 9.8-10.2; NFO: 10 mg, pH: 7.0-8.0.

The values of the parameters of the Langmuir and Feundlich isotherms are shown in Table 6.5. The correlation coefficients (r^2) for both models range between 0.952-1.000, indicating an excellent fit with the observed data. The Langmuir model, with r^2 >0.999 for all three materials, appears to describe the adsorption slightly better than the Freundlich model with a little worse linear correlation (Table6.4).

Table 6.5: Isotherm parameters for adsorption of As(III) on BYC, MHZR and NFO.

Materials	Langmuir			Freundlich		
	r^2	q_m (mg/g)	k_l (L/g)	r^2	n	k_f ((mg/g)(L/mg)$^{1/n}$)
BYC	1.000	34.6	0.303	0.980	1.02	8.43
MHZR	0.999	15.9	0.786	0.960	1.52	0.26
NFO	0.999	16.8	0.505	0.952	1.73	5.94

The saturation capacities of the BYC, MHZR and NFO estimated with the Langmuir isotherm are 34.6 mg/g, 15.9 mg/g and 16.8 mg/g, respectively. The fitting curves also suggest that As(III) was adsorbed in form of a monolayer coverage on the surface of these materials (Langmuir, 1961). All three possess a high adsorption capacity for As(III), but with BYC showing a significantly higher value than the other two. Compared to the results of *Wasay et al.* (1996) and *Suzuki et al.* (2000), the adsorption capacity of BYC and MHZR prepared in this study was somewhat lower. Although the specific surface is the lowest for BYC (6.68 m^2/g), followed by NFO (75.9 m^2/g) and MHZR (373 m^2/g) (Chapter 4), the ranking according to their adsorption capacity is the other way around, with BYC > NFO > MHZR. This clearly shows that the adsorption capacity of these materials does not directly depend on their specific surface. The ranking of BYC and MHZR is in line with the values reported by *Wasay et al.* (1996) and *Suzuki et al.* (2000), with the adsorption capacity of BYC being far larger than that of MHZR. The possible reasons can be a higher density of the functional ion (CO_3^{2-}) on BYC than that (X^-) on MHZR. Meantime, according to these data from Freundlich isotherm, all three of the investigated materials appear to efficiently adsorb As(III) in the considered concentration range (Treybal, 1998).

6.3.3.3 Selective adsorption

It has been reported that adsorption of As(III) on BYC and MHZR is not considerably affected by co-existing anions in solution, including Cl^-, SO_4^{2-} and NO_3^- (Wasay et al., 1996; Suzuki et al., 2000). Adsorption of As(III) on

NFO in the presence of Cl^- and SO_4^{2-} anions was investigated in the optimum pH range (7.0-9.0). Results are presented in Figure 6.7. In the first 15

min, adsorption of As(III) ncreased rapidly to 75.6 %, reaching an equilibrium value of 87.6 % within one hour. This indicates that the NFO prepared in this study adsorbs rapidly As(III) even if Cl^- and SO_4^{2-} coexist in the solution. This is different from the iron-oxide based sorbent reported by Zuo et al. (Zuo et al., 2012), showing that arsenic adsorption was efficient, but significantly affected by the presence of other anions, like Cl^- and SO_4^{2-}. Compared to the other two adsorbents considered in this study, NFO turned out to show the highest adsorption kinetic rate toward As(III). It was shown that the adsorption equilibrium attained within one hour, as compared to 2 h (for MHZR) and 15 h (for BYC) (Figure 6.8).

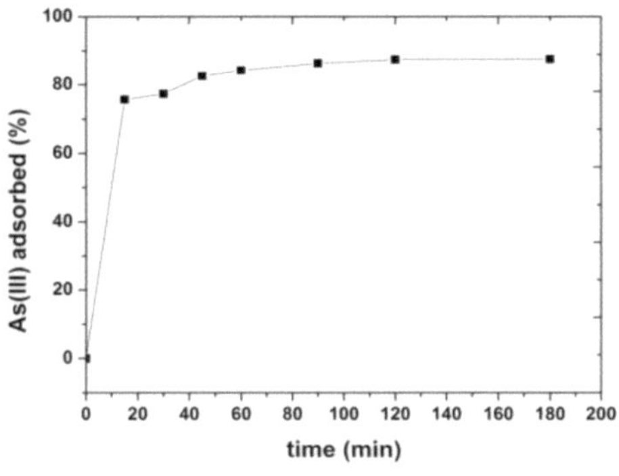

Figure 6.7: Sorption of As(III) on NFO as a function of time. (100 mg NFO in 50 ml solution with 1.5 mg/L As(III), 100 mg/L Cl^- and 100 mg/L SO_4^{2-} solution; stirring rate 150 rpm; pH 7.32).

Changes in pH, and in the concentration of co-existing Cl^- and SO_4^{2-} anions during the near-quantitative adsorption of As(III) on the three materials are shown in Figure 6.8. From a solution with about 1200 µg/L As(III), 95.1 %, 99.3 % and 98.2 % of As(III) was adsorbed within 24h, 3h and 1h, on BYC, MHZR and NFO, respectively. The pH of the solution shaken with MHZR decreased from 9.76 to 8.90 during the experiment, most probably due to the release of H^+ from the surface of the resin to the solution. In the experiment with BYC, the pH also slightly decreased from 9.96 to 9.64, as a consequence of the exchange of CO_3^{2-} in BYC with $H_2AsO_3^-$ in solution. In the NFO experiment, the pH remained practically unchanged, because no ionic surface processes have been involved. No significant change in concentrations of Cl^- and SO_4^{2-} has been recorded during adsorption of As(III) on three of the adsorbents. Most significantly for the purpose of the study, it indicates that all of the tested materials can be used to separate As(III) from Cl^- and SO_4^{2-}, but the best results have been achieved with NFO, which was able to adsorb 98.2 % of As(III) within only 1 hour.

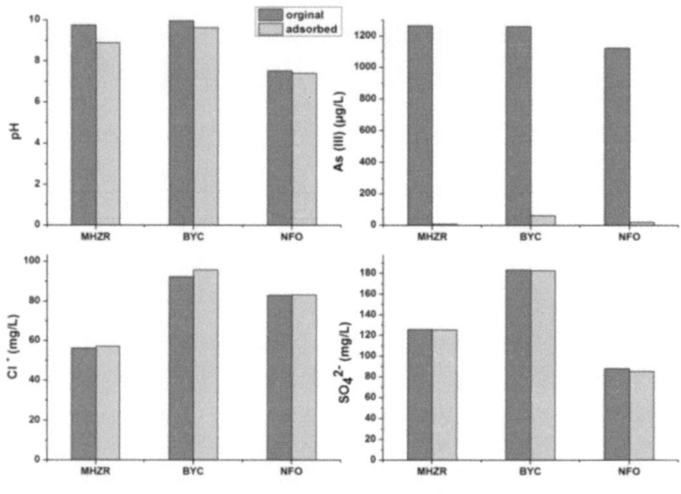

Figure 6.8: Changes in pH and concentrations of As(III), SO_4^{2-} and Cl^- during the nearly quantitative adsorption of As(III) on MHZR (100 mg), BYC (25 mg) and NFO (10 mg). Shaking rate: 150 rpm; reaction times: MHZR 3 hours, BYC 24 hours, NFO 1 hour.

6.3.3.4 Stripping of arsenite from the adsorption materials

The optimal conditions for the stripping of As(III) from the sorbents for the purpose of subsequent precipitation and isotopic measurement was tested by applying different conditions, i.e. desorption agents: 0.05 M HNO_3, 1 M NH_4NO_3 buffer solution, and 0.5 M NaOH (Figure 6.9). Only 0.2 %, 2.0 % and 22.6 %, respectively, of As(III) was released from MHZR, after shaking for 24 hours in 0.05 M HNO_3, 1 M NH_4NO_3 buffer solution, and 0.5 M NaOH, respectively. Considerably better results have been achieved with BYC. With 0.05 M HNO_3, almost all of As(III) (95.0 % of total) was stripped, while only 3.2 % was desorbed under alkaline condition. With 1 M NH_4NO_3 buffer solution, an intermediate recovery (41.3 %) was obtained. From NFO, after shaking for 1 hour with the afore mentioned agents, 8.9 %, 6.9 % and 65.8 % of As(III) was released, with 0.5 NaOH giving the best (but still relative moderate) results.

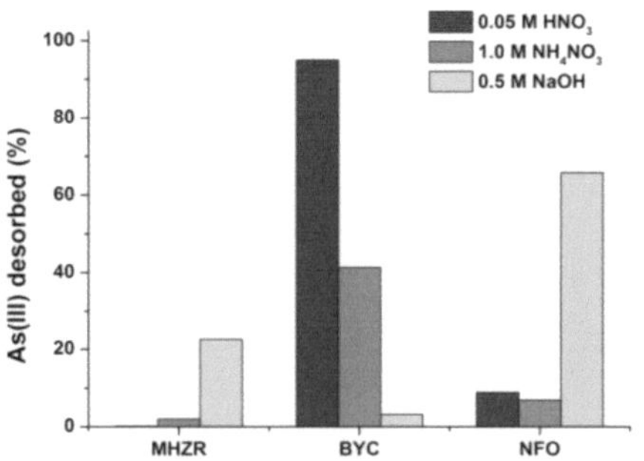

Figure 6.9: Desorption of As(III) by different agents. 20 mL of each agent shaken at a rate of 150 rpm with BYC (25 mg) and NFO (10 mg) for 1 hour, and with MHZR (100 mg) for 24 hours.

In order to optimize desorption, notably for MHZR and NFO, in addition to the above mentioned standard conditions, different concentrations of acid or alkaline agents were also tested. Results for MHZR are shown in Table 6.6. The recovery was not considerably improved, the highest desorption rate of 27.5 % was achieved with 3 M NaOH. In contrast, the desorption from NFO was significantly improved, by applying high concentrations of NaOH (Table 6.7). 85.0 % and 87.5 % of As(III) were desorbed in 50 mL of 1 M and 2 M NaOH solution, respectively. Because the difference between using 1 and 2 M is negligible, and the conditions for AsO_3^{3-} precipitation are close to neutral, the use of 1 M NaOH is the better option to desorb As(III) from NFO. Based on these results, it can be concluded that MHZR is inadequate for the separation and pre-concentration of As(III), and is not considered further.

Table 6.6: Desorption of As(III) from MHZR by using different concentrations of NaOH and HNO_3. Experimental conditions: MHZR: 0.5 g; volume of eluents 20 mL; shaking rate 150 rpm; contact time 1 hour.

Desorption agent	mol/L	As(III) recovery (%)
NaOH	0.05	8.0
	0.25	19.2
	0.5	25.7
	1.0	11.4
	2.0	18.1
	3.0	27.5
HNO_3	0.05	0.2
	0.1	0.1
	0.5	0.3
	1.0	0.1
	2.0	0.3

Table 6.7: Desorption of As(III) from NFO by using different concentrations of NaOH. Experimental conditions: NFO 20 mg: volume of eluents 50 mL; shaking rate 150 rpm; contact time 1 hour.

Desorotion agent	mol/L	As(III) recovery (%)
NaOH	0.01	24.1
	0.05	49.6
	0.1	51.2
	0.5	75.9
	1.0	85.0
	2.0	87.5

6.4 Summary

The experiments for selective and quantitative separation of AsO_3^{3-} from HCO_3^- and SO_4^{2-} were first achieved by using methods of pH adjusting and precipitation with Ba^{2+}, respectively. This developed method is suitable for the separation of As(III) from other oxyanions in water and aqueous solutions, however, during the precipitation of Ag_3AsO_3, the co-occurring Cl^- in water was also precipitated, which leads to an excessive "dilution" of the material used for the isotope measurement. In case of natural water with high Cl^- content, Cl^- must be eliminated prior to precipitation. Therefore, the three different materials (BYC, MHZR and NFO) were applied to selectively adsorb and separate As(III) oxyanions from SO_4^{2-} and Cl^- in water samples. The results showed recoveries for As(III) of 95.1 %, 99.3 % and 98.2 % by BYC, MHZR and NFO, respectively. In the same time, the SO_4^{2-} and Cl^- anions which can perturb the determination of the isotopic composition of oxygen in As(III) oxyanion, are practically completely eliminated. However, 95.0 % and 87.5 % of AsO_3^{3-} can be desorbed from BYC and NFO, but only 27.5 % from MHZR using different desorption agents. Therefore, MHZR is inadequate for the purpose of this study. The results of adsorption with IRAN show that it can adsorb and desorb AsO_3^{3-}, SO_4^{2-} and Cl^- simultaneously, which means that it is unfortunately not suitable for the selective separation of As(III).

The above results show that As(III) can be successfully separated from As(V) and PO_4^{3-} by MAS, while the perturbing anions Cl^-, and SO_4^{2-} can be eliminated subsequently by BYC or NFO. Finally, the AsO_3^{3-} can be almost quantitatively precipitated as Ag_3AsO_3. The combination of these steps into a single process-flow, allowed the elaboration of a comprehensive working protocol for the extraction of AsO_3^{3-} from water for oxygen isotope work, as shown schematically in Figure 6.10. The whole process takes around in the isolated and reducing conditions, which avoid the affection from oxygen in air. The detailed steps for the separation of AsO_3^{3-} can be summarized as follow:

1) First, pH value of the natural sample is adjusted to 4.3 to remove HCO_3^- as CO_2 gas. Then, the adsorbent MAS can separate As(III) from As(V) and PO_4^{3-} in the pH range of 4-9, and As(V) and PO_4^{3-} are adsorbed on MAS. Meantime, the anions like Cl^-, and SO_4^{2-} are together with As(III) in the effluent, as well as SO_4^{2-}, which is partly released from MAS.

2) Second, there are two options to separate As(III) from Cl^- and SO_4^{2-} by using adsorbents BYC and NFO, respectively. During the separation experiment with BYC, the solution containing As(III), Cl^- and SO_4^{2-} is mixed with BYC in the pH range between 9.8 and 10.5 by shaking at a rate of 150 rpm for 15 hours. During this process, only As(III) can be quantitatively adsorbed on BYC, but SO_4^{2-} and Cl^- are still in solution. Then As(III) can be totally desorbed from BYC in 0.25 M HNO_3, in which BYC was dissolved as Y^{3+} and CO_2 gas. As(III) in $AgNO_3$ solution is precipitated as Ag_3AsO_3 by adjusting pH to 7.5. In the separation process with NFO, As(III) is quantitatively separated from SO_4^{2-} and Cl^- by physical adsorption on NFO in the pH range of 7.0-9.0 at a shaking rate of 150 rom for 1 hour. Around 90 % of As(III) on NFO can be desorbed in 1 M NaOH solution and then be quantitatively precipitated as Ag_3AsO_3 by adjusting pH to 7.5.

Although the desorption of As(III) from NFO was little relatively lower than from BYC, the comparison of those two methods for separation of As(III) from natural water shows that NFO is most effective to separate As(III) from other disturbed anions, which takes only 4-5 hours for the whole experimental procedure. Meantime, there is no ion exchange reaction during the adsorption process. Whereas the experimental process by BYC takes around 20 hours and there is ion exchange reaction during adsorption process. The con-existed Y^{3+} with As(III) in the desorption solution can possible change the oxygen values in AsO_3^{3-}, which will show in the further chapter. Therefore, NFO is obviously better than BYC for the separation of As(III) from natural water.

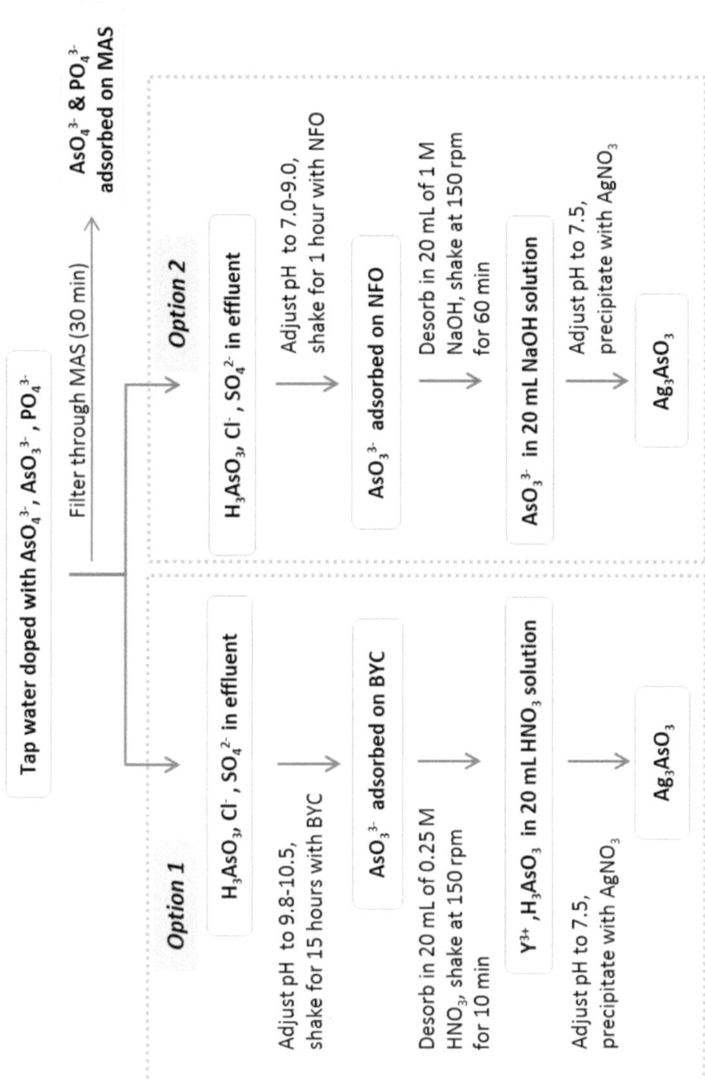

Figure 6.10: Schematic process-flow with two options (two adsorbents) for extraction of As(III) from AsO_3^{3-} contaminated natural water samples. (working temperature: 21±1°C).

7 Measurement of $\delta^{18}O$ values in Ag_3AsO_4 and Ag_3AsO_3

7.1 Introduction

Arsenic is a trace and toxic elements in the environment, occurring principally as AsO_4^{3-} and AsO_3^{3-} with a minor amount of methyl and dimethyl arsenic compounds. One method for the isotope composition of oxygen in inorganic arsenic species (AsO_4^{3-} and AsO_3^{3-}) in natural water was developed in the study. Following the procedure for selective separation of arsenic oxyanions from water, the last step is to extract AsO_4^{3-} and AsO_3^{3-} from water by means of precipitation as a solid compound. The measurement of oxygen isotope composition in the precipitates was afterwards carried out.

There is no detailed method for preparation of arsenic samples and determination of $\delta^{18}O$ values in arsenic species by using modern methods of isotope ratio mass spectrometry. However, the methods for measurement of $\delta^{18}O$ in PO_4^{3-} (O'Neil et al., 1994; Stephan, 2000; Vennemann et al., 2002), SO_4^{2-} (Mangalo et al., 2007) and NO_3^- (Singleton et al., 2005) have been routinely performed by on-line coupling of Thermal Conversion/Element Analyzer (TC/EA) with high temperature reduction with graphite and continuous flow mass spectrometry (IRMS) (TC/EA-Py-CF-IRMS). The preparation of PO_4^{3-} followed Crowson et al. (Crowson et al., 1991), which became the standard method for PO_4^{3-} isotopic work (e.g.,Lecuyer et al., 2004; Stephan, 2000). The procedure is based on the method described by Firshing (1961), by isolating PO_4^{3-} from the pre-concentrated sample solution as Ag_3PO_4. Silver phosphate (Ag_3PO_4) is much better than $BiPO_4$ (Tudge, 1960) for analysis because of its easy preparation and not hygroscopicity. The determination of $\delta^{18}O$ values is based on the TC-EA (high temperature conversion-elemental analyzer)/IRAMS (isotope ratio-mass spectrometry) method. The samples is analyzed by first high temperature reduction of Ag_3PO_4, forming CO as the analyte gas, then passing the CO gaseous products in GC separation and introducing them into the mass spectrometer using He carrier gas (eg, (Kornexl et al., 1999a; Vennemann et al., 2002)). The measurement of $\delta^{18}O$ in AsO_4^{3-} and AsO_3^{3-} in this study consulted the methods that applied for the determination of oxygen isotope in PO_4^{3-}.

Based on the rapid precipitation over a wide pH range without incorporating structural H_2O molecules, Ag^+ was also selected as for quantitatively precipitating AsO_4^{3-} and AsO_3^{3-} as Ag_3AsO_4 and Ag_3AsO_3, which is an important step for collection arsenic oxyanions from water. There is no report for quantitatively precipitation of AsO_4^{3-} and AsO_3^{3-} from solution, but the quantitative precipitation of PO_4^{3-} from homogeneous solution (Firsching, 1961; Lecuyer et al., 2007). This chapter defines a procedure for quantitatively precipitating AsO_4^{3-}/AsO_3^{3-}, based on the description in several reports (Vogel, 1961; Jander und Blasius, 1988). A successful analytical method for measuring the composition of stable oxygen isotope within AsO_4^{3-}/AsO_3^{3-} was also established by using EA-Py-CF-IRMS. Meantime, the kinetic exchange of oxygen isotope between AsO_4^{3-} and water is also presented in this chapter.

7.2 Experiments

7.2.1 Precipitation of silver phosphate (Ag_3PO_4)

Precipitation of PO_4^{3-} as Ag_3PO_4 was first done for confirmation of the method described in Firsching (1961), and also for the precise of measurement of $\delta^{18}O$ values in PO_4^{3-}. This method basically consists of precipitation of Ag_3PO_4 from the pre-concentrated sample solution, making use of the solubility of Ag_3PO_4 in ammoniac solution. 0.0689 g KH_2PO_4 was weighted and dissolved in 100 ml water, which was placed in a 250 mL beaker. About 1 ml of concentrated NH_4OH was added to raise the pH to 9-10. Fifteen ml of buffered silver ammine solution was added to the flask.

The buffered silver ammine included 0.5 M $AgNO_3$, 2.9 M NH_3NO_3 and 0.75 ml concentrations NH_4OH. Upon heating of this solution in a thermostatic bath (70°C), as ammonia volatilizes and pH gradually decreases, millimeter-size yellowish crystals of Ag_3PO_4 started to precipitate at about pH 8.5 from basic solution. The reaction ended at pH 7.5 and the precipitation was quantitative. The crystals were collected on a 0.45 µl Millipore filter, and washed for three times with Mill-Q H_2O, and then air-dried at 50°C for 5-6 hours.

7.2.2 Preparation of internal reference material (NBS 120c)

The protocol isolates PO_4^{3-} from apatite as Ag_3PO_4 crystals using acid dissolution and anion-exchange resin, which is a well-known method for preparation of internal standard Ag_3PO_4. About 225 mg (~50 mg PO_4^{3-}) NBS 120c material (Florida) was weighted and dissolved in 11.2 ml of 2 M HF over 24h. The solution was then centrifuged and separated from the CaF_2. 16.8 ml of 2 M KOH was added to neutralize the solution then placed in a 250 ml Erlenmeyer flask. About 1 ml of 28 % NH_4OH was added to raise the pH of solution to 9-10. Fifteen ml of buffered silver ammine solution was finally added to the flask to precipitate PO_4^{3-}, which followed the same method as section 7.2.1.

7.2.3 Precipitation of the arsenic oxyanions

The tests were processed in 100 mL of solution with 100 mg $Na_2HAsO_4 \cdot 7H_2O$ and 100 mg $NaAsO_2$. Arsenate in solution was precipitated by additional of weak acidic $AgNO_3$ solution (pH: 5.5), then pH value of mixed solution was adjusted to 7.5 by pure HNO_3 and NaOH solutions. To avoid the O-isotope composition exchange between AsO_4^{3-} and ambient water, the process of Ag_3AsO_4 precipitation was rapidly, ca. 5-10 min. The crystals were collected on a 0.45 µl Millipore filter, and washed with deionized H_2O three times, and then air-dried at 50°C for 5 hours. Silver arsenite is prepared in a similar way as the process for Ag_3AsO_4. A series of batch experiments were carried out to precipitate AsO_4^{3-} and AsO_3^{3-} in 50 ml water with arsenic concentration of 0.5 mg/L, 1 mg/L, 5 mg/L, 10 mg/L, 50 mg/L, 100 mg/L, 250 mg/L and 500 mg/L. After adjusting pH to 7.5, the solution was centrifuged for 10 min, Ag^+ in excess was subsequently removed by washing the precipitates with de-ionized water for three times. Finally, ethanol was used to dry the precipitate. All the samples were collected and milled in homogeneous powder, which were measured the oxygen isotope composition to check the stability of oxygen isotope value in the same batches. The storage of Ag_3AsO_3 precipitate must be in vacuum conditions for less than 2 weeks, because it is easily oxidized on air. The mineral structure of Ag_3AsO_4 precipitate was checked by X-ray diffraction (XRD).

7.2.4 Measurement of $\delta^{18}O$ value in solid and water

The $\delta^{18}O$ value was measured in solid and liquid phase by on-line coupling of Thermal Conversion/Element Analyzer (TC/EA) with high temperature reduction with graphite and continuous flow mass spectrometry (IRMS) (TC/EA-Py-CF-IRMS) and Liquid water isotope analyzer (LWIA-24d). The detailed information has been described in section 3.4.5. Each sample was measured at least three times, to ensure the correction of the $\delta^{18}O$ values and the deviation is as small as possible.

7.2.5 Experiments of oxygen exchange between arsenate and water

Solution with 450 mg/L AsO_4^{3-} were prepared in 1000 mL of deionized water, which has a $\delta^{18}O$ value of 12.0 ‰. 300 mL of solution with original pH value of 7.5 was spited up for three portions, whose pH was adjusted to 1.0, 4.6 and 10.2 by using pure 3 mL of 65 % HNO_3, 120 µL of 65 % HNO_3 and 100 µL of 28 % NH_4OH, respectively. The resulting solutions had a same concentration of As(V) but different dominant AsO_4^{3-} species. The species of H_3AsO_4, $H_2AsO_4^-$, and $HAsO_4^{2-}$ are the dominant species in the solution with pH range of 1, 5 and 10, respectively (Linge, 2002). Solutions with different pH were filled in separate vials (14 mL) without space on the top, and then closed tightly with silicate cap. All vials were put in oven at 25 °C and 70 °C. The equilibration time was lasted for 2 weeks. Before precipitation, pH of all samples was adjusted to 9 by using 28 % NH_4OH. 0.5 mL of 1 mol/L $AgNO_3$ was then added for rapid and quantitative precipitation of Ag_3AsO_4. The newly formed precipitates were agitated, collected via centrifugation, washed with deionized water, and finally dried at 70 °C overnight. The total time for precipitation and washing was 50 min, in which 5 min for water sample, 5 min for precipitation, 40 min for washing at 3000 rpm for for 4-5 times. Two parallel trials were done for each condition.

7.3 Results and discussion

7.3.1 Recovery of phosphate, arsenate and arsenite in precipitates

After precipitation of PO_4^{3-} in ammoniac solution with Ag^+ (Firsching 1961; Stephan, 2000), 98.03 % and 94.15 % of PO_4^{3-} were precipitated as Ag_3PO_4 from solution with KH_2PO_4 and NBS120c material, respectively (Table 7.1). The results proved that the method of precipitation with Ag^+ is practical and the similar one can be also applied to precipitate arsenic species in solution. The precipitates Ag_3PO_4 from NBS120c were employed as an internal standard for calibrating the value of oxygen isotope.

The precipitation was based on the property of Ag_3AsO_4/Ag_3AsO_3, to be dissolved by ammoniac as $[Ag(NH_3)_2]^+$ and AsO_3^{3-}/AsO_4^{3-} (Vogel, 1961). In principle, $AgNO_3$ solution is added to the weak acidic sample and subsequently neutralized slowly with ammoniac. Considering the low solubility of Ag_3AsO_4 and Ag_3AsO_3, the most appropriate way to extract arsenic oxyanions from an aquatic solution is their precipitation with $AgNO_3$. This approach is much similar to the method currently used for the precipitation of PO_4^{3-} (Crowsonet al., 1991; Stephan, 2000). The difference was that the precipitation of AsO_4^{3-} and AsO_3^{3-} were carried out more rapidly, which is advantageous in preventing the oxidation of As(III) to As(V), from a mixed solution with AsO_3^{3-} and Ag^+ ions in excess. There was no oxygen fractionation in the PO_4^{3-} during the repadly precipitation as Ag_3PO_4 at pH 7.5 (Dettman et al., 2001), therefore, the quantitative precipitation of Ag_3AsO_4/Ag_3AsO_3 was achieved within 10 minutes by adjusting the pH to 7.5. Those different conditions were tested for the precipitation of AsO_4^{3-} (Table 7.2). In a solution with 450 mg/L AsO_4^{3-}, whose pH value was 5.2, only 62.4 % of AsO_4^{3-} was precipitated as Ag_3AsO_4 without pH change. Whereas, 95.3 % of AsO_4^{3-} was precipitated when pH was rapidly adjusted to 7.5 by using NH_4OH. The results indicate that AsO_4^{3-} can be quantitatively and rapidly precipitated as Ag_3AsO_4 by adjusting pH to 7.5. The same method was also applied to precipitate Ag_3AsO_3 and around 94.2±1.3 % (n=5) of AsO_3^{3-} was precipitated by adjusting pH to 7.5. The results also reveal that pH value is a most important factor for the precipitation process. The replicates for each precipitate are presented in Table 7.3 and the recovery of each material was very stable in the certain conditions.

By adjusting the pH to neutral (pH 7.0-7.5), low to high concentrations of AsO_4^{3-} and AsO_3^{3-} was quantitatively precipitated as Ag_3AsO_4 and Ag_3AsO_3, respectively. The effect of concentrations of AsO_4^{3-}/AsO_3^{3-} on the recovery of precipitation is shown in Table 7.4. It can be seen, that 99 % of AsO_4^{3-} can be recovered as Ag_3AsO_4 when the concentration of As(V) ranged between 0.5 mg/L and 550 mg/L. In contrast, the recovery of As(III) as Ag_3AsO_3 was relatively low, with the concentrations of As(III) between 0.5 mg/L and 14 mg/L. At higher As(III) concentrations (55 mg/L), AsO_3^{3-} recovery is higher. At 55 mg/L, 78.2 % of AsO_3^{3-} was recovered, and around 99 % of AsO_3^{3-} was precipitated with As(III) concentrations > 100 mg/L.

Table 7.1: The first tests for precipitation of Ag_3PO_4. The experiment was carried out in 50 mL of reaction solution.

Chemical	Weighted value (g)	Weighted value (g)
NH_4NO_3	2.0002	3.4817
KH_2PO_4	0.0689	-
NBS120c material	-	0.2294
$AgNO_3$	2.5471	1.2927
0.045 Filter	0.0936	0.0906
0.045 µm filter +Ag_3PO_4	0.2908	0.2939
Ag_3PO_4	0.1972	0.2042
Recovery	98.03 %	94.15 %

Table 7.2: Tests for precipitation of AsO_4^{3-} under different conditions. Experimental conditions: 20 mL batches with pure AsO_4^{3-}; concentration of the added $AgNO_3$ solution: 10 g/L; reaction time: 5 min.

Sample	Agent for pH adjusting	pH	Recovery (%)
A	Without adjusting	5.2	62.4
B	NH_4OH	7.5	95.3

Table 7.3: The duplications of each precipitate under different pH conditions. Experimental conditions: 20 mL batches with pure PO_4^{3-} (270 mg/L), AsO_4^{3-} (450 mg/L) and AsO_3^{3-} (500 mg/L); concentration of the added $AgNO_3$ solution: 10 g/L; reaction time: 5-10 min.

Samples	Agent for pH adjusting	pH	Recovery (%)	Replicates
Ag_3PO_4	NH_4OH	7.5	97.5±0.7	n=2
Ag_3AsO_4	Without adjusting	5.3	64.4±4.8	n=10
Ag_3AsO_4	NH_4OH	7.5	96.5±5.6	n=10
Ag_3AsO_3	NH_4OH	7.5	94.2±1.3	n=5

Table 7.4: Effect of As concentrations on the precipitation of Ag_3AsO_3 and Ag_3AsO_4 with $AgNO_3$ solution. Experimental conditions: 20 mL batches with pure As(III) and As(V); concentration of the added $AgNO_3$ solution: 10 g/L ; reaction time: 5 min; pH adjusted to 7.5 with NH_4OH or HNO_3.

Concentration of As(V) or As(III) (mg/L)	**Recovery of As(V) (%)**	**Recovery of As(III) (%)**
0.5	99.3	11.3
1.5	99.9	8.9
7.0	99.7	15.0
14	100.0	14.3
55	100.0	78.2
110	99.9	99.8
280	99.9	99.3
550	99.3	98.4

7.3.2 Structure identification of silver arsenate

The identities of the precipitates were first confirmed using powder X-Ray diffraction (Figure 7.1). The reflections of Ag_3AsO_4 match standard reflections from powder diffraction files (006-0493), from the International Centre for Diffraction Data, Newtown Square, PA, USA). No $AgNO_3$ reflection appeared within the pattern, which indicats that the precipitate was pure Ag_3AsO_4. The colors of the precipitates are consistent with those reports with the powder diffraction files and literature reports, and offer an additional line of evidence for mineral identification (Vogel. 1961). The color of Ag_3PO_4, Ag_3AsO_4 and Ag_3AsO_3 presented as yellow-green, red-brown and bright yellow, respectively.

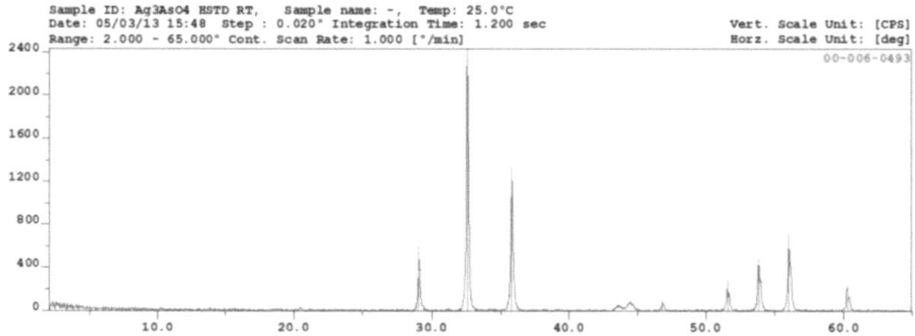

Figure 7.1: X-Ray diffraction (XRD) patterns for AsO_4^{3-} precipitate with Ag^+ (Ag_3AsO_4). Reflections of reference material reparted in powder diffrection cards (PDF) are plotted for comparison. Reference Ag_3AsO_4 in PDF number: 00-006-0493.

7.3.3 Establishment of method for measurement of $\delta^{18}O$ value

To establish a stable method and make an exact calibration for the measurement of $\delta^{18}O$ value are the most efforts in the study. With the method of TC/EA-Py-CF-IRMS (Vennemann et al., 2002; Lécuyer et al, 2007), the system stability was first detected. The mean $\delta^{18}O$ values in the Ag_3PO_4, which precipitated from KH_2PO_4 and NBS120c, were 1.47 ± 0.20 ‰ (n=10) and 23.03 ± 0.18 ‰ (n=6), respectively. The $\delta^{18}O$ value is reproducible and stable, which indicates that the thermal conversion system with 1450 °C can be applied for evaluating the O-isotope in solid phase.

According to the experiences with the method of TC/EA-IRMS from the isotope geochemistry laboratory in University of Lausanne, precipitates like $BaSO_4$ can be partly remained in the bottom of the carbon column with 1450 °C, which could affect the $\delta^{18}O$ value in the following measured samples. Therefore, to avoid the effect of SO_4^{2-}, the standard materials will not use NBS 127 ($BaSO_4$). The $\delta^{18}O$ value in reference pure water (USGS47 and VSMOW2) was first evaluated to avoid any contaminations, and other reference materials are two Ag_3PO_4 reference materials (LK-2 and LK-3), with different weight of oxygen isotope, and also two KNO_3 reference materials (USGS34 and IAEA N3). The $\delta^{18}O$ values of all reference materials are presented in Table 7.5. The results show that the $\delta^{18}O$ values in pure water and Ag_3PO_4, which were determined in laboratory, have a good correlation with the reference value. In parallel the $\delta^{18}O$ value in KNO_3 cannot show the same correlation between the measured values to the reference values. The possible reason may be the existed NO, which was decomposed from KNO_3 during the thermal conversion process. The existed NO can affect the detection precise of CO by mass spectrometry due to the mass of $^{14}N^{16}O$ (30) is same as $^{12}C^{18}O$ (30). NO (30) with the same mass as CO (30) can be detected in the same time by mass spectrometry (MS). Therefore, the reference materials, pure water and Ag_3PO_4, were selected for calibrating the $\delta^{18}O$ values in arsenic oxyanions. The calibration curve was established by collecting all the measured data of the standards materials through the whole measurement (Figure 7.2), in which the correlation coefficient $R^2 = 0.9990$, and the calibration equation is:

$$y = 1.02988\,x - 10.68401 \tag{1}$$

Table 7.5: Comparison of the $\delta^{18}O$ values in measured data and reference data for each reference materials.

Reference material $\delta^{18}O$ (VSMOW)	**Analysis value (‰ VSMOW)**			**Reference value (‰ VSMOW)**
	Average	Deviation	n	
USGS47 (H_2O)	-9.1	0.41	3	-19.8
VSMOW-2 (H_2O)	11.0	0.28	5	0
LK-2 (Ag_3PO_4)	21.9	0.14	8	12.1
LK-3 (Ag_3PO_4)	27.6	0.31	8	17.9
USGS34 (KNO_3)	-20.4	0.82	7	-27.9
IAEA N3 (KNO_3)	31.4	0.65	6	25.6

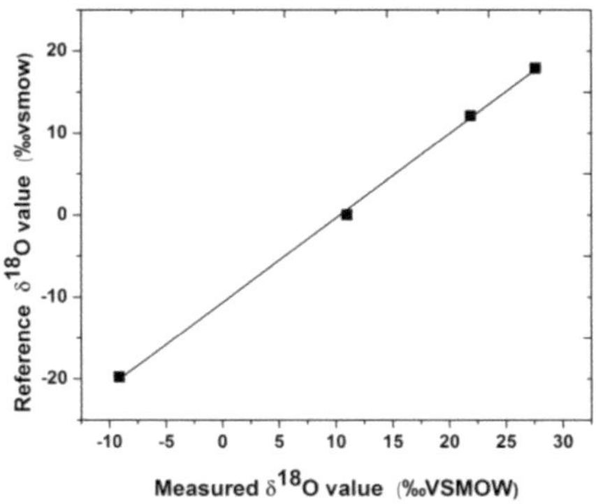

Figure 7.2: The calibration curve for the calculation of $\delta^{18}O$ value (‰ VSMOW). The standard materials: USGS47 (H_2O), VSMOW-2 (H_2O), LK-2 (Ag_3PO_4), LK-3 (Ag_3PO_4).

7.3.4 The analysis of $\delta^{18}O$ value in Ag_3AsO_4/Ag_3AsO_3

By calculating the calibration in Figure 7.2, the comparable mean $\delta^{18}O$ values in Ag_3AsO_4 and Ag_3AsO_3 materials, which prepared from $Na_2HAsO_4 \cdot 7H_2O$ and $NaAsO_2$ at different pH, were evaluated (Table 7.6). The internal standard "AsStd" (pure Ag_3AsO_4) was prepared from 100 mL of solution with 2 g/L dissolved As(V), which was for testing the stability of the $\delta^{18}O$ value in arsenic samples during the measurements. The $\delta^{18}O$ value in "AsStd" was very stable and with a small standard deviation (<0.3 ‰) by measuring with TC/EA-IRMS. The spectrum curve for measuring "AsStd" is shown in Figure 7.3. The total measuring time for one sample was 800 seconds, and the $\delta^{18}O$ value in CO reference gas was measured three times at beginning and also one more time after the measurement of sample, which monitored the stable of each measurement. The stable of $\delta^{18}O$ value and $\delta^{13}C/^{12}C$ in CO reference gas guaranteed the precise of $\delta^{18}O$ value in samples. The measuring time of O isotope in sample was in 350-480 seconds, and the peak center was on 388.26 second. Then, 0.5-0.8 mg of "AsStd" was measured during each 20 times of samples, to control the whole measuring system. The results show that $\delta^{18}O$ value in "AsStd" was very stable, as -3.98 ± 0.27 ‰ (n=19) during the 300 measurements.

The $\delta^{18}O$ values in precipitate, either Ag_3AsO_4 or Ag_3AsO_3, were obviously different from the original materials, $Na_2HAsO_4 \cdot 7H_2O$ or $NaAsO_2$. However, the $\delta^{18}O$ values in H_2O before and after precipitation process were no difference (Table 7.7). The reason for the difference in $\delta^{18}O$ values between the original solids and the precipitates was that the existence of crystal H_2O in $Na_2HAsO_4 \cdot 7H_2O$, whereas $NaAsO_2$ captured or exchanged oxygen atom from H_2O to form AsO_3^{3-}. Compare to the large volume of water for dissolving AsO_4^{3-} and AsO_3^{3-}, the effect of the dissolved process for the $\delta^{18}O$ values in H_2O can be ignored.

The $\delta^{18}O$ values in old samples were different from the fresh samples (around 2 ‰ difference in two years). The $\delta^{18}O$ values in Ag_3AsO_4 could be probably changed in open air with time, which means that arsenic samples for internal standards must be checked in time, and better way is to prepare fresh one before measuring. The novel phenomenon is that the $\delta^{18}O$ value in Ag_3AsO_4 and Ag_3AsO_3 can be a big difference under the different pH for precipitation (Table 7.6). For each batchs with different pH (each of them with 4 duplicates), the pH values were increased by 28 % NH_4OH. When the precipitation of Ag_3AsO_4 under pH value at 7.5, the $\delta^{18}O$ value was -5.34 ± 0.51 ‰ (sample As5_7, n=12). Whereas, the $\delta^{18}O$ value in Ag_3AsO_4, which was precipitated in solution with pH 9.5, was -0.06 ± 0.48 ‰ (sample As5_9, n=12). Meantime, the $\delta^{18}O$ value in Ag_3AsO_3 was -11.9 ± 0.19 ‰ (sample As3_7, n=12) and -14.2 ± 0.28 ‰ (sample As3_9, n=12), which were precipitated under pH value 7.5 and 9.5, respectively. There was 5.2 ‰ of the $\delta^{18}O$ value lower in the batch of As5_7 than in the batch of As5_9. Whereas, there was 3.3 ‰ of the $\delta^{18}O$ value in the batch of As3_7 higher than in the batch of As3_9. The reason for the big differences between the samples only with different pH value is unclear. The possible reasons could be two: 1) the precipitation of AsO_4^{3-} or AsO_3^{3-} were not quantitatively when pH value was 9.5, because Ag^+ can combine with NH_4^+ as $[Ag(NH_3)_2]^+$ when the pH value was adjusted to 9.5 by high concentration of NH_4OH in the precipitation solution (Vogel, 1961). During the uncompleted precipitation process at pH 9.5, the total information of $^{18}O/^{16}O$ in AsO_4^{3-} may be lost in the precipitate Ag_3AsO_4. For example, part of AsO_4^{3-}

with heavy oxygen molecular ^{18}O could still remain in solution, because AsO_4^{3-} with light molecular like ^{16}O might probably be prior to precipitate with Ag^+ as Ag_3AsO_4 (Cole & Chakraborty, 2001). 2) the kinetic oxygen fractionation in AsO_4^{3-} or AsO_3^{3-} might happen very quickly and unstably during equilibrium dissociation reaction like $H_2AsO_4^- \leftarrow\rightarrow H^+ + HAsO_4^{2-}$ system or $H_3AsO_3 \leftarrow\rightarrow H^+ + H_2AsO_3^-$ as pH changed from 7.5 to 9.5. It is not easy to find the real reason to explain the oxygen isotope's behavior in AsO_4^{3-}/AsO_3^{3-} and water systems in so short time (10-30 min). The $\delta^{18}O$ values in Ag_3AsO_4 and Ag_3AsO_3 depended strongly on the precipitation conditions, especially pH value of the reaction solution.

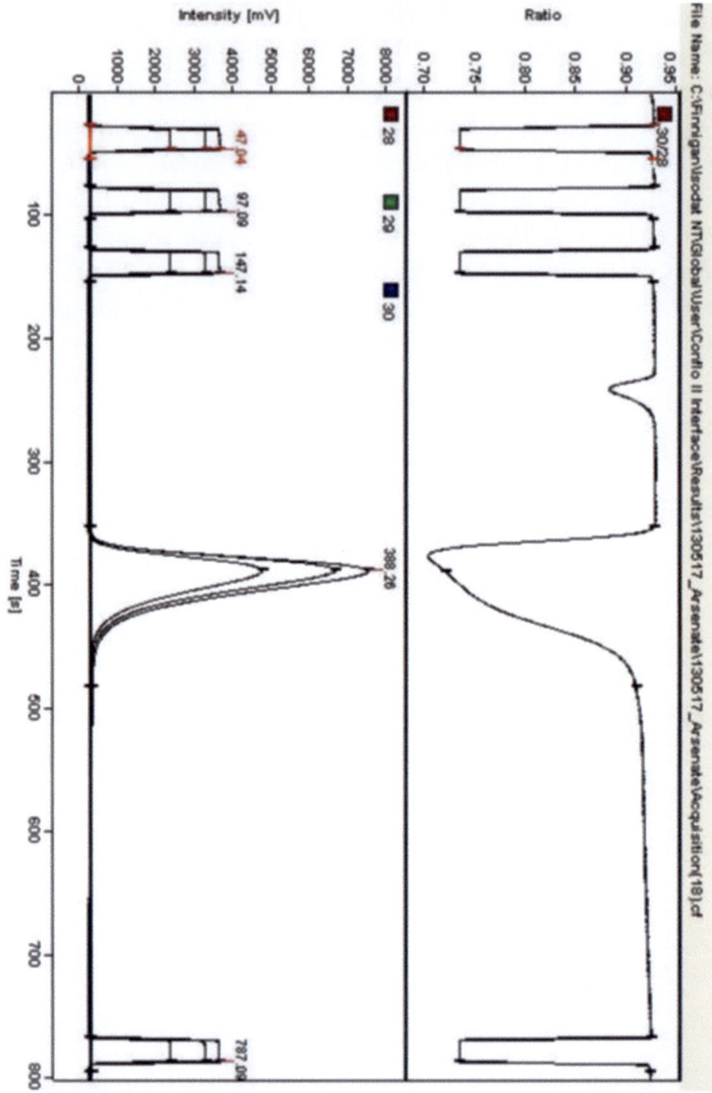

Figure 7.3: The spectrum curve of the detection of oxygen isotope in Ag_3AsO_4, sample: 813 µg, detection time 800 seconds, thermal conversion: 1450 °C, GC: 70 °C.

Table 7.6: Measurement of the $\delta^{18}O$ value in Ag_3AsO_4 and Ag_3AsO_3 with TCEA-IRMS. The internal standard "AsStd" was prepared in 100mL with solution with 2 g/L As(V) at pH 5.2. Samples of As5_7 and As5_9 (Ag_3AsO_4) were taken from the homogenous solution with 1 g/L As(V) but precipitated under pH value at 7.5 and 9.5, which were adjusted by 7 µl and 25 µL 28 % NH_4OH, respectively. For precipitation of As3_7 and As3_9 (Ag_3AsO_3), the operation was the same as As3_7 and As3_9.

Samples	pH for precipi-tation	Analyzed $\delta^{18}O$ value (‰VSMOW)			Comparable value (‰VSMOW)	Note
		Average	Stdev	n		
AsStd (Ag_3AsO_4)	5.2	6.50	0.27	19	-3.98	1 duplicate Prepared in Aug. 2010
As5_5 (Ag_3AsO_4)	5.2	6.48	0.48	15	-4.00	5 duplicates Prepared in Aug. 2010
As5_N (Ag_3AsO_4)	7.5	7.25	0.14	15	-3.20	5 duplicates Prepared in Sept. 2010
As5_7 (Ag_3AsO_4)	7.5	5.18	0.51	12	-5.34	4 duplicates Prepared in April 2013
As5_9 (Ag_3AsO_4)	9.5	10.3	0.48	12	-0.06	4 duplicates Prepared in April 2013
As3_7 (Ag_3AsO_3)	7.5	-1.23	0.19	12	-11.9	4duplicates Prepared in April 2013
As3_9 (Ag_3AsO_3)	9.5	-3.45	0.28	12	-14.2	4 duplicates Prepared in April 2013

Table 7.7: The δ18O values in solid and liquid phase before and after precipitation. The precipitation were done in 20 mL of solution with 200mg/L As(V) and 500 mg/L As(V), with 1.2 mL of 0.5 M $AgNO_3$ for 5 min, and 7 µl of 28 % NH_4OH for pH adjusting, pH for precipitation:7.5.

Samples	δ18O value in solid (‰VSMOW)			δ18O value in H_2O (‰VSMOW)			Note
	Average	Stdev	n	Average	Stdev	n	
As5_0	5.37	0.35	3	-7.98	0.00	2	$Na_2HAsO_4{\cdot}7H_2O$ dried at 50°C for 4 hours; pure water
As5_1	-4.79	0.28	3	-7.92	0.02	2	Ag_3AsO_4; water with As(V)
As5_2	-5.04	0.27	3	-8.00	0.03	2	Ag_3AsO_4; water with As(V)
As5_3	-5.07	0.11	3	-7.93	0.03	2	Ag_3AsO_4; water with As(V)
As5_4	-5.06	0.40	3	-8.02	0.04	2	Ag_3AsO_4; water with As(V)
As3_0	2.51	0.00	3	-7.90	0.09	2	$NaAsO_2$ dried at 50°C for 4 hours
As3_1	-7.57	0.21	3	-7.85	0.11	2	Ag_3AsO_3; water with As(III)
As3_2	-7.56	0.10	3	-7.84	0.08	2	Ag_3AsO_3; water with As(III)
As3_3	-8.14	0.04	3	-7.93	0.08	2	Ag_3AsO_3; water with As(III)
As3_4	-8.17	0.10	3	-7.87	0.03	2	Ag_3AsO_3; water with As(III)
As3_5	-8.44	0.05	3	-7.91	0.05	2	Ag_3AsO_3; water with As(III)

7.3.5 Kinetics of exchange between arsenate and water

The measured δ18O values of dissolved AsO_4^{3-} for all kinetics experiments are presented in Figure 7.4 as a function of different pH at different temperatures. The δ18O value in H_2O was not measureable change during the runs because of the large ratios of water to AsO_4^{3-}. The δ18O value in AsO_4^{3-} decreased rapidly to a new steady state with increasing reaction time. The relative slower decrease in AsO_4^{3-}-δ18O values was observed in the experiments at lower temperature (25 °C) and higher pH conditions (pH=10)

(Figure 7.4). At a low temperature (25 °C), a dissolved AsO_4^{3-} with initial $\delta^{18}O$ value of 1.51 ‰ reached to new steady-state values of -0.40 ‰, -3.40 ‰ in 12 hours and 3 hours in the solution with pH value of 1 and 5, respectively. The new equilibrium of the $\delta^{18}O$ value in dissolved AsO_4^{3-} (7.86 ‰) was achieved at 25 °C and pH 10 after 312 hours, which is much slower comparing to other conditions. At the relative high temperature (70 °C), the new steady-state values were promptly achieved in 3 hours at low or high pH conditions, which were -3.73 ‰, -7.58 ‰ and -11.4 ‰ at pH 1, pH 5 and pH 10, respectively (Table 7.9).

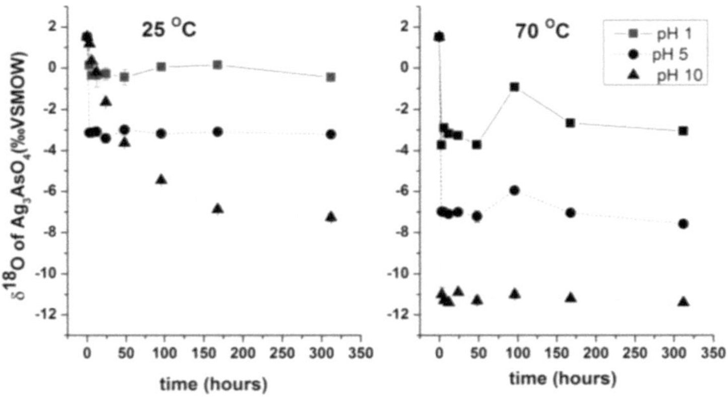

Figure 7.4: Variation with time of the oxygen isotope composition of dissolved AsO_4^{3-} ($[AsO_4^{3-}]$: 450 mg/L) for pH ranging from 1 to 10 under 25 °C and 70 °C. The initial $\delta^{18}O$ value of H_2O was -12.0 ‰.

The change rate of oxygen isotope composition for a first-order isotopic exchange reaction of oxygen between AsO_4^{3-} and water is described by equation 7.1 (Cole and Chakraborty, 2001):

$$f = 1 - e^{-kt} = \frac{\delta^{18}O_i - \delta^{18}O_t}{\delta^{18}O_i - \delta^{18}O_e} \qquad (7.1)$$

where f is the fraction of isotopic exchange between AsO_4^{3-} and water, k is the reaction rate constant for oxygen isotope exchange between AsO_4^{3-} and water (h^{-1}), $\delta^{18}O_i$ is the initial oxygen isotope composition of AsO_4^{3-} at t=0,

$\delta^{18}O_t$ is the composition measured at time t, $\delta^{18}O_e$ is its composition at equilibrium. The diagrams for the half-lives of the each reaction in the experiments are shown in Figure 7.5 and Table 7.8. The negative slopes of the straight lines in Figure 7.5 are the rate constants k of the reaction for pH value of 1, 5, 10 at 25 °C and 70 °C, respectively.

The As-O bond must be broken during the oxygen isotope exchange between AsO_4^{3-} and water and the activation energy required for the oxygen isotope exchange is dependent upon AsO_4^{3-} speciation. In general, the activation energy of a reaction (E_a) is expressed by the Arrhenius equation (Lasaga, 1998):

$$k = Ae^{-(\frac{E_a}{RT})} \tag{7.2}$$

Then

$$\log k = \log A - \frac{E_a}{R} \cdot \frac{1}{T} \tag{7.3}$$

where A is the pre-exponential factor, E_a is the activation energy, R is the ideal gas constant (R = 8.3144621(75) J K^{-1} mol^{-1} (Mohr et al., 2008) and T represents the absolute temperature(K). Values of *log k*, which were calculated from Figure 7.4 (Table 7.9), are plotted vs *1/T* in an Arrhenius plot in Figure 7.6. Because of the quick exchange between AsO_4^{3-} and water, there are only 2 temperature points for each diagram, which can calculate rough activation energies (E_a) from the slopes of the plot in different pH, which are also given in Figure 7.6.

According to the calculation method for kinetic fractionation (Lecuyer et al., 1999; Northrop & Clayton, 1966), the equilibrium of oxygen isotope fractionation factor between AsO_4^{3-} and water (α) are calculated by the equation (7.4), in which n is the total number of mole of ^{18}O or ^{16}O in AsO_4^{3-}. The equilibrium $\delta^{18}O$ values of AsO_4^{3-} (‰VSMOW) and *1000ln* α values are shown in Table 7.9. The relationships between the obtained 1000ln α values and temperature or pH are shown in Figure 7.7 and Figure

7.8, which show a strong linear correlation with for a give pH and temperature.

$$\alpha = \frac{n^{18}O_{arsenate}/n^{16}O_{arsenate}}{n^{18}O_{water}/n^{16}O_{water}} = \frac{1000\ \delta^{18}O_{arsenate}+1}{1000\ \delta^{18}O_{water}+1} \qquad (7.4)$$

The relationship between the half-lives (ln 0.5) of the oxygen isotope exchange reaction for AsO_4^{3-}-water vs pH at each temperature are shown in Table 7.8 for data from this study, showing that the reaction trend is quite similar to the previous studies (Larese-Casanova & Blake, 2012; Okumura & Okazaki, 1973), although the exchange process in this study was much faster than that in those two studies.

At 25 °C, the fractionation expression is exactly for H_3AsO_4-H_2O, $H_2AsO_4^-$-H_2O, and $HAsO_4^{2-}$-H_2O exchange because H_3AsO_4, $H_2AsO_4^-$, and $HAsO_4^{2-}$ are the dominant species at pH =1, 5 and 10, respectively (Linge, 2002). The observed complete oxygen isotope fractionation in $H_2AsO_4^-$-H_2O ($t_{1/2}$= 0.71 hour) was showed much faster than in $HAsO_4^{2-}$-H_2O ($t_{1/2}$ = 38.5 hour) in the study (Table 7.8). Although AsO_4^{3-} and PO_4^{3-} have so similar chemical characteristics, the kinetics of the oxygen exchange reaction in dissolved AsO_4^{3-}-water is much faster than that of dissolved PO_4^{3-}-water (Okumura et al., 1995). There are 10^7 magnitudes of the rate of oxygen fractionation higher in $H_2AsO_4^-$-H_2O systems than in $H_2PO_4^-$-H_2O systems at 30 °C (Kumura et al. 1995). It was calculated as 1×10^{-4} s^{-1}, 1.2×10^{-5} s^{-1}, 7.4×10^{-2} M^{-1} s^{-1} (I (ion strengths)= 0.55 mol/L, 30 °C), for the rate of oxygen exchange in $H_2AsO_4^-$-H_2O, $HAsO_4^{2-}$-H_2O, and $H_2AsO_4^-$-$HAsO_4^{2-}$, respectively (Kumura et al. 1995). As the data shown above, the oxygen exchange is even faster in groundwaters, typically with pH range of 7 - 8.5 and $H_2AsO_4^-$ and/or $HAsO_4^{2-}$ being the dominant species.

Largely similar to other oxyanions-water systems, the isotope exchange rate between dissolved AsO_4^{3-} and water was strongly dependent on temperature and solution pH (Beck et al., 2005; Chiba & Sakai, 1985; Kaneko & Poulson, 2013; O'Neil et al., 2003). The exchange between AsO_4^{3-} and water is faster at low pH (value of 1 and 5) than at high pH (value of 10), but the fastest oxygen change was observed at pH 5, where $H_2AsO_4^-$ is the dominant species (Table 7.9). It is little different from that in PO_4^{3-}-water systems (O'Neil et al., 2003), or in the analogous SO_4^{2-} and water systems (Mizutani & Rafter, 1969), which is much faster at low pH than at high pH. However, the trend of the results in the study is similar to the report for the oxygen exchange between AsO_4^{3-} and water in Kumura et al. (1973). Like description in O'Neil et al. (2003) that the heavy water systems are much farther removed from equilibrium than the light water system, the exchange rates between AsO_4^{3-} and water at 25 °C are uniformly higher in heavy water than in light water (data was not shown).

The exchange rate is very rapid in this study and the concentration of AsO_4^{3-} affects the exchange rate of oxygen isotope between AsO_4^{3-} and water. The results from *Larese-Casanova & Blake* (2013) showed that the complete exchange between AsO_4^{3-} and oxygen was achieved in 2 weeks, which may be caused by the high concentration (30 mM AsO_4^{3-}). All Exchange rates decreased with increasing concentration of PO_4^{3-} in the PO_4^{3-}-water systems (O'Neil et al., 2003). The concentration of AsO_4^{3-} also affects the reaction energy. For example, E_a for an oxygen exchange between $H_2ASO_4^-$ and H_2O in solution with 0.07 mol/L (~9.72 g/L AsO_4^{3-}) AsO_4^{3-} was 28 kJ/mol (Kumura et al. 1995), but that was 4.25 kJ/mol in the reaction of this study, which was running in solution with 200 mg/L As(V) (Figure 7.6). The results mean that the reaction energy is needed lower for oxygen isotope exchange in solution with low concentration, and further indicate that the kinetic exchange of oxygen isotope is faster in low concentration than in high concentration of AsO_4^{3-}.

The species of aqueous arsenate that are dominant in low-pH solutions are enriched in ^{18}O relative to those that dominate at high pH at both 25 °C and 70 °C (Figure 7.7). From the systematics observed, it appears that the order of ^{18}O enrichment among arsenate species at a given temperature is $H_3AsO_4 > H_2AsO_4^- > HAsO_4^{2-} > AsO_4^{3-}$, which is similar to what have been observed in some other oxyanions-water systems, such as phosphate-water systems (O'Neil et al., 2003) and sulfate-water systems (Chiba & Sakai, 1985), carbonate-system (Beck et al., 2005; McCrea, 1950; Usdowski & Hoefs, 1993). It shows that the presence of protons on oxyanions appears to affect profoundly the tendency to concentrate heavy isotopes (O'Neil et al, 2003). The different ionic charges, ion-water interactions, and the degree of dissociation of the various species could be the reasons leading to the different rate of isotope exchange among those species in solution. Figure 7.8 gives a rough trend that the rate of oxygen exchange in AsO_4^{3-}-water systems at low temperature is slower than high temperature at all pH ranges, which is also similar to the trend in PO_4^{3-}-water systems (Lecuyer et al., 1999; O'Neil et al, 2003).

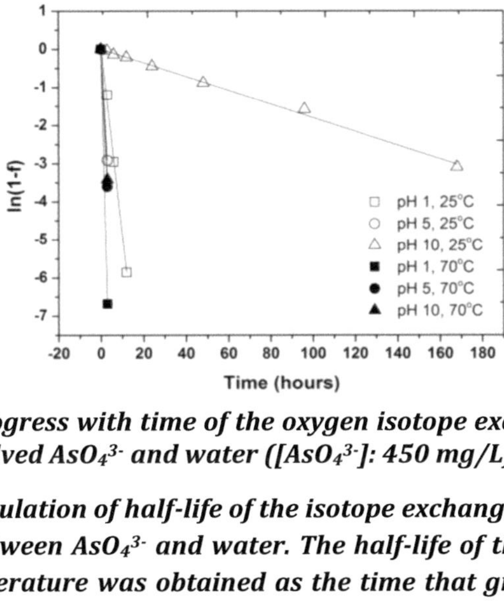

Figure 7.5: Progress with time of the oxygen isotope exchange reaction between dissolved AsO_4^{3-} and water ([AsO_4^{3-}]: 450 mg/L).

Table 7.8: Calculation of half-life of the isotope exchange and fractionation factor between AsO_4^{3-} and water. The half-life of the reaction, $t_{1/2}$, for each temperature was obtained as the time that gives ln (1-f) = ln 0.5. f: the fraction of isotope exchange between AsO_4^{3-} and water.

Temperature (°C)	pH	Equation	R^2	$t_{1/2}$ (hour)
25	1	ln(1-f) = -0.496 t	0.998	1.4
	5	ln(1-f) = -0.972 t	1	0.71
	10	ln(1-f) = -0.018 t	0.998	38.5
70	1	ln(1-f) = -2.229 t	1	0.31
	5	ln(1-f) = -1.202 t	1	0.58
	10	ln(1-f) = -1.139 t	1	0.61

Table 7.9: Calculated rate constants (h^{-1}), equilibrium $\delta^{18}O$ values of dissolved AsO_4^{3-} (‰VSMOW), and fractionation factors between AsO_4^{3-} and water in the pH range of 1-10.

Temperature (°C)	pH	log *k*	$\delta^{18}O$ (AsO_4^{3-}) (when t→∞)	1000 ln ⍰ (AsO_4-H_2O)
25	1	-0.305	-0.45	11.7
	5	-0.012	-3.40	8.95
	10	-1.75	-7.86	4.77
70	1	0.348	-3.73	8.34
	5	-0.042	-7.58	5.07
	10	0.057	-11.4	1.01

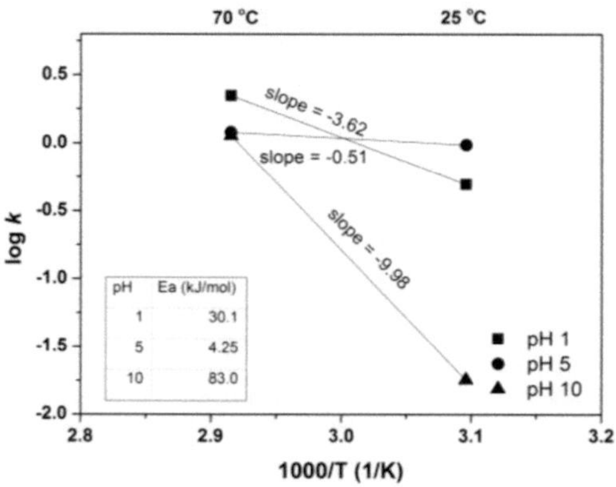

Figure 7.6: Arrhenius plots of oxygen exchange reaction between dissolved AsO_4^{3-} and water ([AsO_4^{3-} = 450 mg/L] by calculating log k vs 1/temperature (1/K). the activation energy Ea is given by the slope of the linear equation.

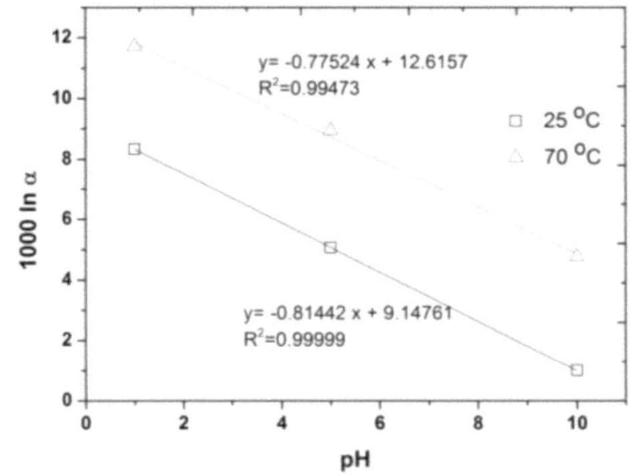

Figure 7.7: Oxygen fractionation between dissolved AsO_4^{3-} and water under different temperatures as a function of pH ($[AsO_4^{3-}]$ = 450 mg/L).

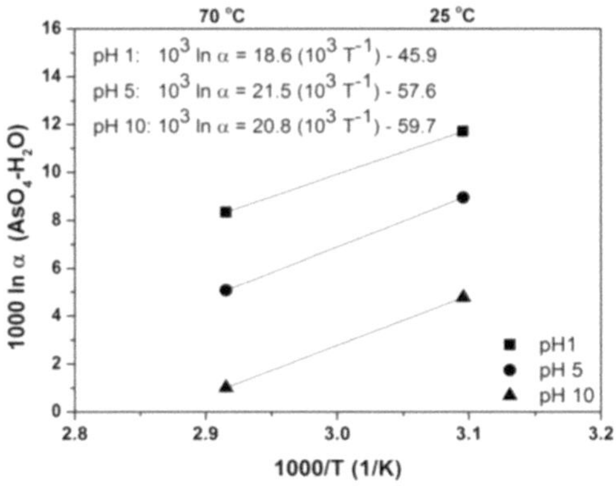

Figure 7.8: Oxygen isotope fractionation between dissolved inorganic AsO_4^{3-} and water in the range 25-70 °C ($[AsO_4^{3-}]$ =450 mg/L).

7.4 Summary

The method for evaluation of the $\delta^{18}O$ values in Ag_3AsO_4 and Ag_3AsO_3 was successfully established in this study. The detailed conditions for precipitation of AsO_4^{3-} and AsO_3^{3-} as Ag_3AsO_4 and Ag_3AsO_3 were clearly specified. The quantitative precipitation of them needs to adjust pH by 28 % NH_4OH to neutral condition, around 7.5. The recoveries of AsO_4^{3-} and AsO_3^{3-} from solution in Ag_3AsO_4 and Ag_3AsO_3 were more than 95 %. The Ag_3AsO_3 precipitate, which is easy oxidized on air, was cooling dried in vacuum drier and stored in vacuum desiccator for a while. The XRD result shows that the precipitates (Ag_3AsO_4) were pure enough for the detection of oxygen isotope composition.

The measurement of the $\delta^{18}O$ value in arsenic oxyanions was done by using Thermal Conversion Elemental Analyzer with isotope ratio mass spectrometry (TC/EA-IRMS). The measured data provided reproducible and the standard deviation about 0.27 ‰ for 19 duplicated samples during 4 days' measurements. The $\delta^{18}O$ values in arsenic oxyanions were very reproducible under different precipitation pH conditions. However, there was quite large difference among the samples, which were precipitated at different pH. Those differences reveal that the pH value of the precipitation solution is very important for the stable and corrected $\delta^{18}O$ value in AsO_4^{3-}/AsO_3^{3-}. There is difference in the $\delta^{18}O$ values of the solid phase of Ag_3AsO_4 samples, which were prepared in the same conditions but different times (old sample were prepared in August 2010, and new samples were prepared in April 2013). These results indicate that the $\delta^{18}O$ values in Ag3AsO4 could slowly change in the long term, around 2 ‰ in 2.5 years.

The rate of oxygen isotope exchange between AsO_4^{3-} and water was investigated at 25 °C and 70 °C and pH 1 to 10. Oxygen exchange proceeded as a first-order reaction, and the exchange rate was strongly affected by temperature and pH, with increased rates of isotope exchange at higher temperature, and lower pH. The exchange rate from fast to slow at different pH conditions was: pH5 >pH 1>pH 10. The complete exchange happened in 3 hours at 25 °C and pH 5, as well at 70 °C and pH range of 1-10. The lowest exchange rate was at 25 °C and pH 10, and the half- time for isotope exchange was 38.5 hours. Arsenate speciation has a significant effect on the order of ^{18}O enrichment among arsenate species at a given temperature, being H_3AsO_4> $H_2AsO_4^-$> $HAsO_4^{2-}$> AsO_4^{3-}. However, the fast exchange rate between AsO_4^{3-} and water at most conditions demonstrates that it is hard to use AsO_4^{3-}-$\delta^{18}O$ to investigate the geochemical behavior of dissolved arsenic oxyanions in environmental conditions.

8 Conclusive discussions

The study has primarily successfully developed all procedures for separation of arsenic oxyanions from natural water. The measurement for the oxygen isotope in arsenic-oxyanions was also reproducible and comparable. The whole procedures for separation of arsenic species from other coexisting anions can establish a complete method for extracting AsO_4^{3-} or AsO_3^{3-} from natural sample for measuring the oxygen isotope composition. Since the kinetics of the oxygen exchange between AsO_4^{3-} and water is too rapid, there may be some oxygenic fractionation in dissolved AsO_4^{3-} or AsO_3^{3-} during the whole separation process. Therefore, the fractionation factor of oxygen isotope in dissolved AsO_4^{3-} or AsO_3^{3-} will be identified and discussed by measurement of $\delta^{18}O$ value before and after the separation procedures in this chapter.

8.1 Fractionation of oxygen isotope in separation procedure of arsenate

As the procedures for separation of AsO_4^{3-} shown in Figure 5.16, the study developed a method for separation of AsO_4^{3-} from AsO_3^{3-} and PO_4^{3-} and three possibilities to separate AsO_4^{3-} from SO_4^{2-} and Cl^-. In those three options, the time for the whole procedure is the same, and around 6 hours. There are always three adsorbents, which must be used in the whole procedure of separation, in which MAS and IRAN are the necessary adsorbents. 1 M base like NaOH can be employed for desorption of AsO_4^{3-} from MAS. The adsorbent IRAN is the only adsorbent, which was used in this study for simple and fast separating AsO_4^{3-} from PO_4^{3-}. The process of separation of AsO_4^{3-} from PO_4^{3-} takes only 2 hours, and high concentration of HNO_3 (2 mol/L) must be used to desorb AsO_4^{3-} but not PO_4^{3-} from IRAN. All the three adsorbents (HMZR, BYC and NFO) for separation of AsO_4^{3-} from SO_4^{2-} and Cl^- have large adsorption capacities for AsO_4^{3-}. Arsenate can be released from those three adsorbents in 1 M base solution like NaOH. The whole process for separation of AsO_4^{3-} from natural samples needs to change the pH of the solution several times, in which the fractionation of oxygen isotope in AsO_4^{3-} must be clarified.

The determination of $\delta^{18}O$ value in the final Ag_3AsO_4 was carried out after the main part of the complete separation procedure with adsorbents IRAN and HMZR. There are big changes of pH (from -0.3 to 14) during the separation experiments with IRAN and HMZR, in which the fractionation of oxygen isotope in dissolved AsO_4^{3-}, was identified. For collection of enough Ag_3AsO_4, the experiment was done in a composite water sample with additional of a high concentration of arsenic, containing 100 mg/L As(V) and 10 mg/L PO_4^{3-}, as well as the major perturbing anions SO_4^{2-} and Cl^-, whose concentration are already shown in Table 4.6. The results in each step of the procedure are summarized in Table 8.1. It shows that 72.3 % of dissolved AsO_4^{3-} was recovered in Ag_3AsO_4, and the detection with EDX-SEM confirms the successful separation (Figure 8.1). The shape/morphology of the grains is quite heterogeneous, and no other elements but only C, O, As, Cl and Ag were identified in the precipitates. The omni-present C peak is a background value which comes from the tape on which the sample was fixed. Whereas the small peak of Cl, may result from the trace elements, which existed in HNO_3 or NH_4OH solution, used to adjust the pH. The same elements have been detected also in the precipitate, which obtained directly from a solution containing only AsO_4^{3-}. There was 8 ‰ difference of $\delta^{18}O$ value between the sample before and after the separation procedure, which means that there was a quite a large fractionation during the procedures. However, the fractionation values of oxygen isotope in dissolved AsO_4^{3-} in those four repeatd of the whole separation procedure are very close , which mean that the separation method of AsO_4^{3-} could be applied to extract AsO_4^{3-} from waters for detection of the $\delta^{18}O$ value in dissolved AsO_4^{3-} with a corrected factor.

The question is that which factors cause the oxygen isotope fractionation happening during the ion exchange process of the separation experiment. The effect of high alkali condition (1 M NaOH) for the fractionation of oxygen isotope between AsO_4^{3-} and water is shown in Table 8.2. There was only 0.3 ‰ differences between the samples, which were directly precipitated from neutral solution, and the samples, which were precipitated from 1M NaOH by adjusting pH to neutral with 65 % HNO_3. This indicates that there is no fractionation between AsO_4^{3-} and water in the alkali solution, which can be applied for desorption of AsO_4^{3-} from adsorbents during the separation procedure.

The preparation procedures for isotope measurements can affect the isotopic fractionations. For example, the process of the phosphoric acid digestion, which is applied to digest carbonate minerals for clumped isotope measurements (Carothers et al., 1988; McCrea, 1950), affected the fractionation of $^{18}O/^{16}O$ by calculating with a theoretical model (Guo et al., 2009). During the process of ion exchange on IRAN and MHZR, AsO_4^{3-} was adsorbed by reacting with function group on the ion exchanger, like OH^- in IRAN and protonated Zr-OH_2^+ in MHZR, and the pH for the reaction with IRAN was 7.5 and with MHZR were 10.0. Those ion exchange processes could be a factor to lead the oxygen isotope fractionation happening ((Maréchal & Albarède, 2002). Another possible reason is the use of 2 M HNO_3, which could lead the AsO_4^{3-} species suddenly changed and the composition of oxygen isotope in AsO_4^{3-} could be restructured during the change of pH (section7.3.4). Although there is no report to mention whether ion-exchanger or strong acid (like HF) could affect the oxygen isotope fractionation during separation of PO_4^{3-} from natural samples ((Crowson et al., 1991; O'Neil et al., 1994; Stephan, 2000), different pH values in solutions have been proven to affect the fractionation of oxygen isotope in different dissolved oxyanions ((Dietzel et al., 2009; Ohmoto & Lasaga, 1982; O'Neil et al., 2003). On other hand, the 6 hours experimental time is also one reason for the fractionation in AsO_4, because of the rapid exchange of oxygen isotope between AsO_4^{3-} and water (section7.3.5). The lost of 34 % of AsO_4^{3-} in the separation step could be also one factor for the change of the $\delta^{18}O$ value in Ag_3AsO_4 precipitate.

Table 8.1: Arsenate recovery and changes in the concentration of coexisting anions, during a complete extraction experiment with IRAN and HMZR.The replicates: n = 4.

Anions		As(V)	SO_4^{2-}	Cl^-	NO_3^-	HCO_3^-	PO_4^{3-}
Initial solution		100 mg/L	67.8 mg/L	36.0 mg/L	4.08 mg/L	342 mg/L	10 mg/L
Solution after adsorption on IRAN		<0.8 mg/L	<0.04 mg/L	<0.06 mg/L	<0.06 mg/L	Not measured	<0.08 mg/L
Desorbed from IRAN in 2 M HNO_3		91.2 ±1.6 %	81.2 ±2.3 %	99.5 ±1.7 %	98.6 ±3.4 %	-	-
Adsorbed on MNZR		96.4 ±0.7 %	-	-	-	-	-
Desorbed from MHZR by 1 M NaOH		76.3 ±1.4 %	-	-	-	-	-
Recovered in Ag_3AsO_4		98.5 ±0.1 %	-	-	-	-	-
Recovered after the whole separation procedure		66.1 ±1.4 %	-	-	-	-	-
$\delta^{18}O$ value (VSMOW) in Ag_3AsO_4	**From pure AsO_4^{3-} solution**	2.50 ±0.28 ‰	-	-	-	-	-
	From separation process	10.5 ±0.08 ‰	-	-	-	-	-

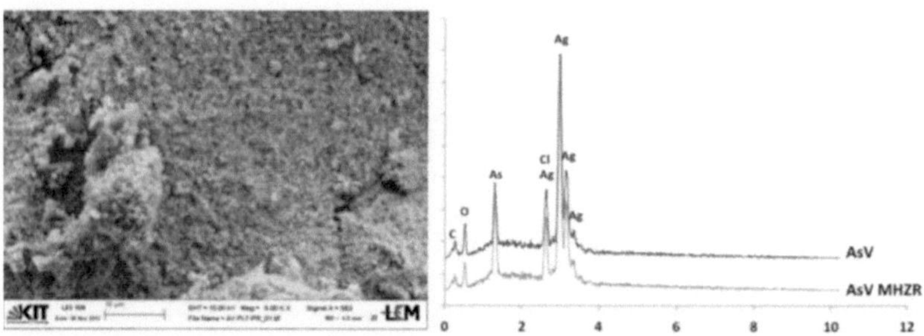

Figure 8.1: SEM micrographs and EDX-spectra of grains obtained by using MHZR (AsIII MHZR) for separation. For comparison the same is shown for Ag_3AsO_4 precipitated from pure As(V) solution (AsV).

Table 8.2: The $\delta^{18}O$ value in dissolved AsO_4^{3-} in different solutions. Samples AsV and AsVNa were Ag_3AsO_4 precipitates but from different solutions, which are mentioned in tables.

Sample name	$\delta^{18}O$ value (VSMOW) in Ag_3AsO_4	Note
AsV	2.81±0.16 ‰ (n=6)	From neutral solution with pH value of 7.5
AsVNa	2.51±0.13 ‰ (n=3)	From the solution with 1M NaOH, pH was adjusted to 7.5 by 65 % HNO_3

Note: 50 mL of solutions with 450 mg/L AsO_4^{3-} were precipitated at room temperature (21±1 °C) in 10 min.

8.2 Fractionation of oxygen isotope in separation procedure of arsenite

As shown schematically in Figure 6.10, the combination of these steps into a single process-flow demonstrates allowed the elaboration of a comprehensive working protocol for the extraction of AsO_3^{3-} from water for oxygen isotope work. The whole procedures of separation need about 16 hours and 3 hours by using BYC and NFO, respectively. The simple and fast procedure for the separation of AsO_3^{3-} from water by first being successfully separated from AsO_4^{3-} and PO_4^{3-} by MAS, while the perturbing anions Cl^-, SO_4^{2-} can be eliminated subsequently by BYC or NFO, and finally the AsO_3^{3-} were almost quantitatively precipitated as Ag_3AsO_3. The feasibility of the procedure was verified by extracting AsO_3^{3-} in form of Ag_3AsO_3 from a composite water sample, containing AsO_4^{3-} and AsO_3^{3-}, as well as the major perturbing anions PO_4^{3-}, SO_4^{2-} and Cl^- (Table 8.3). To ensure no As(III) oxided to As(V) in the whole separation process, All of those experiments were carried out under a closed bottles and all the solution were pumped with N_2 gas to release the dissolved O_2 gas.

The results are summarized in Table 8.3. As shown, AsO_3^{3-}, Cl^-, and SO_4^{2-} remained still in solution after both AsO_4^{3-} and PO_4^{3-} were retained to 100 % in 4 g of MAS. Subsequently, under optimized conditions, 83.2 ± 0.52 % (n=4) and 93.3 ± 2.27 % (n=4) of AsO_3^{3-} were loaded onto 0.25 g of BYC and 0.4 g of NFO, respectively, while Cl^-, and SO_4^{2-} were left behind in the filtered solution. Using 25 mL of 0.25 M HNO_3, 98.1 ± 2.61 % (n=4) of AsO_3^{3-} was stripped almost quantitatively from BYC. Alternatively, 86.9 ± 4.84 % (n=4) of the AsO_3^{3-} loaded onto the NFO was desorbed with 50 mL of 0.5 M NaOH. Finally, the pH of the solutions, containing only the desorbed AsO_3^{3-}, was adjusted with diluted HNO_3 or NH_4OH buffer solutions to neutral or slight acid values, and by adding $AgNO_3$ solution in excess, Ag_3AsO_3 was precipitated at pH 7.5. The precipitates (a light purple precipitate from the extraction with BYC, and a bright yellow one from the separation with NFO) were separated by centrifugation, subsequently thoroughly washed with de-ionized water and dried. The amount of the AsO_3^{3-} recovered in this step was 98.3 ± 1.28 % (n=4) of the AsO_3^{3-} loaded onto the BYC and 95.7 ± 4.81 % (n=4) of that adsorbed on the NFO. The general recovery in form of Ag_3AsO_3 of the whole procedure was 80.2 ± 2.5 % and 77.6 ± 6.5 %, relative to the initial amount of AsO_3^{3-} in the water sample, via separation with BYC and NFO, respectively.

Finally, the purity and elemental composition of the precipitates, which were obtained after passing all steps of the separation procedure, was characterized by SEM-EDX (Figure 8.2). The shape/morphology of the grains is quite similar to the Ag_3AsO_4 precipitates, which got from the separation procedure for AsO_4^{3-}. The precipitate, resulting from the procedure using BYC as adsorbent, contains also Y in addition to the above elements, possibly representing a combination of both Ag_3AsO_3 and traces of $YAsO_3$. Nevertheless, neither Cl nor Y affects the measurements of the isotopic composition of oxygen. Therefore, both options can be used to separate and extract AsO_3^{3-} from natural water samples. The all-over recovery of the whole separation procedure by BYC and NFO is (80.2 ± 2.5 % and 77.6 ± 6.5 %) of the AsO_3^{3-} present initially in solution. The purity of the final Ag_3AsO_3 precipitate is sufficient to be used for determination of the $\delta^{18}O$ value of AsO_3^{3-}.

Although the recovery of AsO_3^{3-} is higher in the separation procedure with BYC, the measurement of $\delta^{18}O$ values in the mix precipitates of Ag_3AsO_3 and $YAsO_3$ is not stable by measuring with TC/EA-IRMS (Table 8.3). It is difficult to find a real reason, which may explain, in part, that available Y^{3+} affects the reactions in the carbon column in TC/EA at high temperature (1450 °C). There were 1.5 ‰ differences of the $\delta^{18}O$ values between Ag_3AsO_3, which were precipitated from solution with pure AsO_3^{3-}, and Ag_3AsO_3, which were extracted from solution after separation procedures with NFO. Compared to the fractionation of oxygen isotope in the separation procedure of AsO_4^{3-}, 1.5 ‰ differences is fairly smaller. The relative shorter reaction time (3 hours) and no addition of strong acids may be the possible reason for the relative small fractionation. Despite of this, 1.5 ‰ difference is still a large gap for a separation method of geochemical work onsite. However, the fractionation values of oxygen isotope in dissolved AsO_3^{3-} in those four repeatd of the whole separation procedure are very close, which mean that the separation method of AsO_3^{3-} could be applied to extract AsO_4^{3-} from waters for measurement of the $\delta^{18}O$ value in AsO_3^{3-} with a corrected factor.

Table 8.3: Arsenite recovery and changes in the concentration of coexisting anions, during a complete extraction experiment.

Anions		**As(III)**	**SO_4^{2-}**	**Cl^-**	**HNO_3^-**	**HCO_3^-**	**PO_4^{3-}**	**As(V)**
Initial solution (mg/L)		100	67.8	36.0	4.08	342	5.01	5
Filtrate after adsorption on MAS (mg/L)		100	113	22.8	3.57	263	<0.01	<0.8
Adsorbed	**On BYC**	83.2±0.5 % (n=4)	-	-	-	-	-	-
	On NFO	93.3±2.3 % (n=4)	-	-	-	-	-	-
Desorbed	**From BYC in 0.05 M HNO_3**	98.1±2.6 % (n=4)	-	-	-	-	-	-
	From NFO in 1 M NaOH	86.9±4.8 % (n=4)	-	-	-	-	-	-
Recovered in Ag_3AsO_3	**From BYC**	98.3±1.3 % (n=4)	-	-	-	-	-	-
	From NFO	95.7±4.8 % (n=4)	-	-	-	-	-	-
Recovered in whole steps	**From BYC**	80.2±2.5 % (n=4)	-	-	-	-	-	-
	From NFO	77.6±6.5 % (n=4)	-	-	-	-	-	-
$\delta^{18}O$ value (VSMOW) in Ag_3AsO_3	**From pure AsO_3^{3-} solution**	1.6±0.27 ‰ (n=3)	-	-	-	-	-	-
	From BYC	Values not stable	-	-	-	-	-	-
	From NFO	0.06±0.47 ‰ (n=4)	-	-	-	-	-	-

Note: Repeated 4 times for each batch of BYC and NFO; "-" indicates not detected

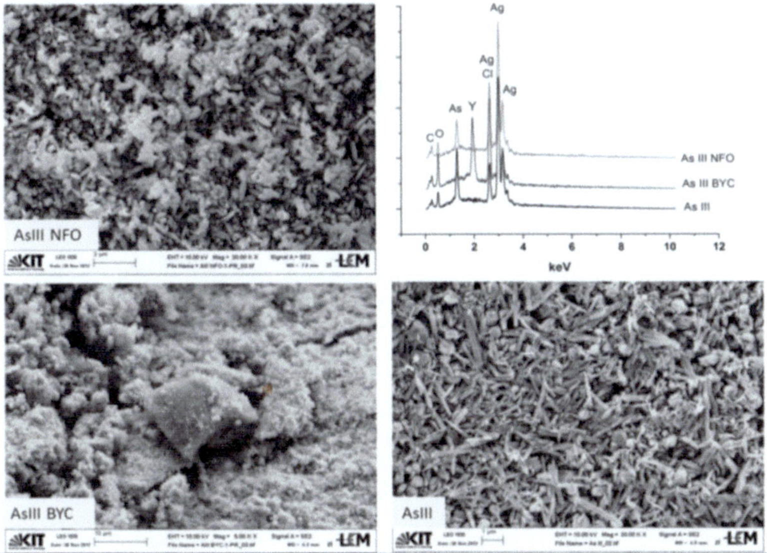

Figure 8.2: SEM micrographs and EDX-spectra of grains obtained by using BYC (AsIII BYC) ans NFO (AsIII NFO) for separation. For comparison the same is shown for Ag_3AsO_3 precipitated from pure As(III) solution (AsIII).

8.3 Discussion

The whole separation procedure of dissolved AsO_4^{3-} and AsO_3^{3-} was successful and with a good recovery of them from the separation eluent as Ag_3AsO_4/Ag_3AsO_3. As the results described before, the minimum amount of Ag_3AsO_4 or Ag_3AsO_3 for the $\delta^{18}O$ value measurement with three time duplicates is 1.5 - 2.4 mg, which means that 0.43-0.69 mg of AsO_4^{3-} or AsO_3^{3-} are required from natural water. Since the As concentrations in natural waters is large range, from less than 0.5 μg/L to more than 5000 μg/L (Smedley & Kinniburgh, 2002), the required volumes of natural water are big difference among different areas. It is hard to collect enough arsenic-species from groundwater in the low concentration of arsenic, which is much too low to permit the efficient extraction of a sufficient amount of

AsO_4^{3-}/AsO_3^{3-} for isotope analysis. In the groundwater with a very high concentration of arsenic (> 500μg/L), the required volume of water is around 1L, which indicates that the rapid collection can be achieved on site. However, the ratio of As(V) and As(III) in groundwater is about 1/4, which means that AsO_4^{3-} is much less than AsO_3^{3-} in groundwater, even in the hotspot areas (Guo et al., 2008b). If extracting enough samples from groundwater with arsenic concentration high 500 μg/L, it is needed 4L of groundwater for AsO_4^{3-} and 1L for AsO_3^{3-}. As the results show in Chapter 7, the concentration of AsO_3^{3-} affects the recovery of AsO_3^{3-} in Ag_3AsO_3, furthermore, the quantitative precipitation of AsO_4^{3-} or AsO_3^{3-} from the constant volume of water is very important for analysis of $\delta^{18}O$ value (Cole & Chakraborty, 2001). Therefore, for collection of enough AsO_4^{3-} and AsO_3^{3-} from natural water, a large volume of water samples must be required for the further analysis of $\delta^{18}O$ value.

Comparison of the adsorbents for the separation experiments, most of the mechanisms for the adsorption of AsO_4^{3-} are ion exchange reaction, which could lead the fractionation of oxygen isotope ((Maréchal & Albarède, 2002). The adsorption on nanoparticle Fe_2O_3 (NFO) is however the process of physisorption and much repaid, which could lend less isotopic fractionation in the separation procedures. Therefore, to shorten separation process and reduce oxygen isotopic fractionation, NFO is the optimal material for the separation of AsO_4^{3-} or AsO_3^{3-} from SO_4^{2-} and Cl^-. Thus, the complete and detailed separation procedure of AsO_4^{3-} and AsO_3^{3-} can be structured as Figure 8.3, in which most of AsO_4^{3-} and AsO_3^{3-} can be extracted from natural samples.

It has been clearly known that isotopic exchange between PO_4^{3-} and water can be rapidly promoted by enzyme-catalyzed reactions ((Blake et al., 1997; Tudge, 1960), that may affect the oxygen isotope composition of fossils preserved in soils and sediments. Oxygen exchange at low temperature between inorganic aqueous solutions and dissolved PO_4^{3-} is very slow, which can be negligible even over geological time scales (e.g. (Kolodny et al., 1983; Tudge, 1960)). However, the significant oxygen exchange can occur between dissolved PO_4^{3-} and water in few weeks at high temperature

(Lecuyer et al., 1999). As a compound with similar chemical characteristics like PO_4^{3-}, the kinetic exchange of oxygen isotope between AsO_4^{3-} and water shows a big fractionation happening in very short time under different temperature and pH values. This gives obvious information that it is a difficult task to use oxygen isotope in arsenic oxyanions as a tool for study the source of arsenic in groundwater. The fractionation of oxygen isotope occurred in the separation process of both AsO_4^{3-} and AsO_3^{3-}, which was developed in this study. It is not only because of the short exchange time between dissolved AsO_4^{3-}/AsO_3^{3-} and water, but also possible because of the adjustment of pH during the separation process, as well the reaction in anion exchange process.

Nevertheless, the fractionation between the adsorbed AsO_4^{3-}/AsO_3^{3-} on minerals and water is still unclear, which may give us some new hints for the arsenic behavior in natural environment. For this issues, many minerals, like Fe/Mn oxides, which are ubiquitous in sediments and interact strongly with dissolved AsO_4^{3-}/AsO_3^{3-} via sorption, co-precipitation, mineral transformation and redox-cycling reaction (Bang & Meng, 2004; Bowell, 1994), must be taken into account for the fractionation of oxygen isotope in adsorbed AsO_4^{3-} and AsO_3^{3-}. Those minerals could be also the important reservoir for dissolved AsO_4^{3-} and AsO_3^{3-}. Arsenate and AsO_3^{3-} bound to those minerals could reflect dissolved AsO_4^{3-} and AsO_3^{3-} sources as well as carrying a history of the biogeochemical cycling of arsenic. Jaisi et al. (2010) reported that oxygen isotope fractionation between adsorbed and aqueous PO_4^{3-} occurs during the early phase of sorption with isotopically-light phosphate ($P^{16}O_4^{3-}$) preferentially incorporated into sorbed/solid phases. Many studies have been reported for the mineral-water systems, like in carbonate-H_2O, layer silicate-H_2O systems, showing that the exchange rates and lattice energies are well correlated (Cole, 2000). The nature of the dissolved mineral species significantly affects fractionations in mineral-H_2O systems (Brandriss, O'Neil, Edlund, & Stoermer, 1998; Cole & Chakraborty, 2001). Taking calcite as an example, which was proven very close to oxygen isotopic equilibrium in the calcite-water system at low temperatures during the precipitation process from a solution (Horita & Clayton, 2007; Zhou, 2005; Zhou & Zheng, 2001). However, many arguments still exist for isotopic fractionation in the calcite-water system, for example, whether or not isotope salt affect of dissolved minerals would occur on oxygen isotpe exchange between water and the interest minerals, or the mineral-water correction is still necessary for calculation of fractionations in mineral-water systems (Horita & Clayton, 2007; Zheng, 1999; Zheng & Zhou, 2007; Zhou, 2005; Zhou & Zheng, 2001).

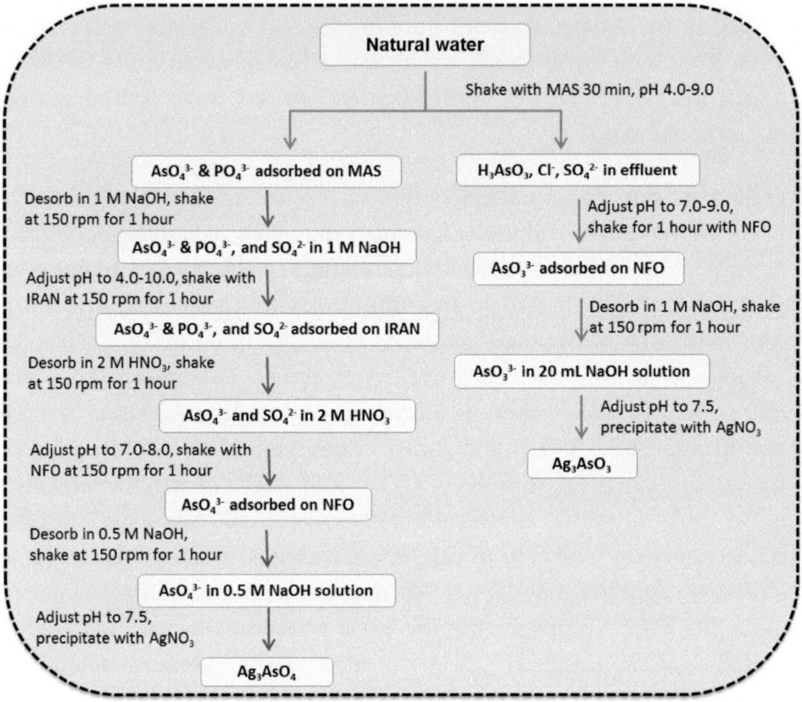

Figure 8.3: The flowchart for the separation of arsenic oxyanions from natural water.

9 Conclusions and outlook

9.1 Conclusions

The experiments conducted in this study show complete methods for separation of AsO_4^{3-} and AsO_3^{3-} from natural water. In natural water, HCO_3^- is first removed as CO_2 by adjusting pH below 4.3. Arsenate and PO_4^{3-} can be then quantitatively separated from AsO_3^{3-}, SO_4^{2-} and Cl^- by an alumino-silicate based sorbent (MAS). The separation of AsO_4^{3-} from PO_4^{3-} was a challenging task due to their similar chemical behavior, which was solved by using Arberlite IRA-400(OH) (IRAN). It can separate 99 % of AsO_4^{3-} from PO_4^{3-}. Subsequencetly, AsO_4^{3-} is quantitatively separated from SO_4^{2-} and Cl^- by using porous polymer beads loaded with monoclinic hydrous zirconium oxide (MHZR), Basic Yutrium Carbonate (BYC), or nano-Fe_2O_3 particle (NFO); whereas, AsO_3^{3-} can be separated from them only with BYC and NFO. During those tests, NFO is identified as the most effective adsorbent because of its adsorption properties. Arsenate and AsO_3^{3-} can be whereafter quantitatively precipitated from separation eluent as Ag_3AsO_4/Ag_3AsO_3 using ammonia to rapidly adjust the pH to 7.5. The $\delta^{18}O$ values in Ag_3AsO_4 and Ag_3AsO_3 are stable and reproducible, which can be calibrated using the international standards of USGS47 (H_2O), VSMOW-2 (H_2O), and also two internal standards of LK-2 (Ag_3PO_4) and LK-3 (Ag_3PO_4), by analyzing with Thermal Conversion/Elemental Analyzer-Gas Chromatogrphy-isotope ratio mass Spectrometry (TC/EA-GC -IRMS).

The trend of kinetic exchange of oxygen isotope between dissolved AsO_4^{3-} and water is similar to that has been observed in some other oxyanions-water systems, such as PO_4^{3-}-water systems (O'Neil et al., 2003), SO_4^{2-}-water systems (Mizutani & Rafter, 1969), or CO_3^{2-}-system (Beck et al., 2005). It is strongly dependent on temperature and pH of solution. The complete exchange of oxygen isotope between AsO_4^{3-} and water is very fast under laboratory temperature and pH conditions. It indicates that oxygen exchange between AsO_4^{3-}/AsO_3^{3-} and groundwater onsite is even more rapidly to a new equilibrium because of the dominating arsenic species $HAsO_4^{2-}$ and $H_2AsO_4^-$ at pH range of 7-8.5. The results from lab experiments confirmed that big fractionation occurred during the process of separation of AsO_4^{3-}/AsO_3^{3-} from natural waters. The fractionation of oxygen isotope in dissolved AsO_4^{3-} or AsO_3^{3-} was very stable, was 8 ‰VSMOW and 1.5 ‰VSMOW, respectively, in the four repeats of the whole separation procedures. Desipte oxygen isotope composition of dissolved AsO_4^{3-} is hard to be a useful tracer to determine the source of arsenic in groundwater of the As-richened areas, the method of separation could be applied to study the adsorbed arsenic oxyanions in minerals of sediments in aquifers.

9.2 Outlook

The study developed several methods for separation of AsO_4^{3-} and AsO_3^{3-} from natural water, which however are hard to be applied for source studies of AsO_4^{3-} and AsO_3^{3-} in fieldwork because of the fast kinetic exchange of oxygen isotope between AsO_4^{3-} and water in aquifer. However, those methods may be used for quantitative extraction of anions, like arsenic, PO_4^{3-}, SO_4^{2-} and Cl^-, from natural water for some other analysis, like speciation. On the other hand, it could be also used for other trace oxyanions from natural water, for which it has been confirmed that the kinetic exchange of oxygen isotope between dissolved phase and water is very slow, like selenite, selenate (Kaneko & Poulson, 2013; Larese-Casanova & Blake, 2012).

The fractionation of oxygen isotope in adsorbed AsO_4^{3-} or AsO_3^{3-} on Fe/Mn –minerals in aquifer must be clear in further work. It must be clear that the exchange of the oxygen isotope in adsorbed arsenic anions with water or other oxyanions, as well as the affection of other compounds in sediment to the fractionation of oxygen isotope. If there is a slow kinetic exchange, the separation process with short operation time, which has been developed in this study, can be applied to inverstigate the geochemical behavior of adsorbed arsenic in aquifer. Furthermore, the development of a method for leaching the adsorbed arsenic oxyanion from sediment to detect $\delta^{18}O$ value is the next important question for natural samples.

The possible reasons for the fractionation of oxygen isotope in AsO_4^{3-} /AsO_3^{3-} during the separation processes give some hints for the analysis of isotope in natural samples. The additional acid and ion exchange reaction could affect the fractionation of oxygen isotope during the process of sample treatment. Therefore, it is necessary to clarify the isotopic fractionation in the extraction procedures of other oxyanions from natural samples, like PO_4^{3-} , CO_3^{2-}, and SO_4^{2-} in liquid or solid phases, must guarantee the analysis precise of oxygen isotope.

References

Acharyya, S.K., Chakraborty, P., Lahiri, S., Raymahashay, B.C., Guha, S., Bhowmik, A. (1999). Arsenic poisoning in the Ganges delta. Nature, 401(6753), 545–discussion 546–7.

Ahn, J.S., Chon, C.M., Moon, H.S., Kim, K.W. (2003). Arsenic removal using steel manufacturing byproducts as permeable reactive materials in mine tailing containment systems. Water Research, 37(10), 2478–2488.

Aiuppa, A., D'Alessandro, W., Federico, C., Palumbo, B., Valenza, M. (2003). The aquatic geochemistry of arsenic in volcanic groundwaters from southern Italy. Applied geochemistry, 18(9), 1283–1296.

Ajioka J. 2012. An economical and easy-to-use biosensor could reduce the chance of being poisoned by arsenic – a common contaminant of wells in parts of Asia. http://www.cam.ac.uk/research/news/new-test-to-detect-arsenic-contamination-in-drinking-water.

Akpomie, G.K., Ogbu, I.C., Osunkunle, A.A., Abuh, M.A., Abonyi, M.N. (2012). Equilibrium isotherm studies on the sorption of Pb (II) from solution by Ehandiagu clay. Journal of Emerging Trends in Engineering and Applied Sciences, 3(2), 354–358.

Appelo, C.A.J., Van Der Weiden, M.J.J., Tournassat, C., Charlet, L. (2002). Surface complexation of ferrous iron and carbonate on ferrihydrite and the mobilization of arsenic. Environmental Science & Technology, 36(14), 3096–3103.

WHO. (2001). Arsenic Compounds, Environmental Health Criteria 224, 2nd ed., Geneva :World Health Organisation.

Audi, G., Bersillon, O., Blachot, J., Wapstra, A.H. (2003). The Nubase evaluation of nuclear and decay properties. Nuclear Physics A, 729(1), 3–128.

Balaji, T., Yokoyama, T., Matsunaga, H. (2005). Adsorption and removal of As(V) and As(III) using Zr-loaded lysine diacetic acid chelating resin. Chemosphere, 59(8), 1169–1174.

Bandyopadhyay, R.K. (2002). Hydrochemistry of arsenic in Nadia District, West Bengal. Journal of the geological society of India, 59(1), 33-46.

Bang, S., Meng, X. (2004). A review of arsenic interactions with anions and iron hydroxides. Environmental engineering research-seoul, 9(4), 184-192.

Beck, W.C., Grossman, E.L., Morse, J.W. (2005). Experimental studies of oxygen isotope fractionation in the carbonic acid system at 15°, 25°, and 40°C. Geochimica et Cosmochimica Acta, 69(14), 3493–3503.

Bednar, A.J., Garbarino, J.R., Burkhardt, M.R., Ranville, J.F., & Wildeman, T.R. (2004). Field and laboratory arsenic speciation methods and their application to natural-water analysis. Water Research, 38(2), 355–364.

Belcher, R., Ingram, G., Majer, J.R. (1969). Direct determination of oxygen in organic materials-I A study of the carbon reduction method. Talanta, 16(7), 881–892.

Berg, M., Tran, H.C., Nguyen, T.C., Pham, H.V., Schertenleib, R., Giger, W. (2001). Arsenic Contamination of Groundwater and Drinking Water in Vietnam: A Human Health Threat. Environmental Science & Technology, 35(13), 2621–2626.

Bhattacharyya, R., Chatterjee, D., Nath, B., Jana, J., Jacks, G., Vahter, M. (2003). High arsenic groundwater: mobilization, metabolism and mitigation--an overview in the Bengal Delta Plain. Molecular and cellular biochemistry, 253(1-2), 347–355.

Bibi, M.H., Ahmed, F., Ishiga, H. (2008). Mobility of arsenic and trace element inventories in sediment cores from Masuda City, southwestern Japan. Environmental Geology, 54(4), 791–803.

Bindeman, I. N., Lundstrom, C. C., Bopp, C., Huang, F. (2013). Stable isotope fractionation by thermal diffusion through partially molten wet and dry silicate rocks. Earth and Planetary Science Letters, 365, 51–62.

Blake, R.E. (2005). Biogeochemical cycling of phosphorus: Insights from oxygen isotope effects of phosphoenzymes. American Journal of Science, 305(6-8), 596–620.

Blake, R.E., Alt, J.C., Martini, A.M. (2001). Oxygen isotope ratios of PO_4: an inorganic indicator of enzymatic activity and P metabolism and a new biomarker in the search for life. Proceedings of the National Academy of Sciences of the United States of America, 98(5), 2148–2153.

Blake, R.E., O'Neil, J.R., Garcia, G.A. (1997). Oxygen isotope systematics of biologically mediated reactions of phosphate: I. Microbial degradation of organophosphorus compounds. Geochimica et Cosmochimica Acta, 61(20), 4411–4422.

Blendell, J.E., Bowen, H.K., Coble, R.L. (1984). High purity alumina by controlled precipitation from aluminum sulfate solutions.American Ceramic Society Bulletin, 63, 797-802

Borgo, C.A., Gushikem, Y. (2002). Zirconium Phosphate Dispersed on a Cellulose Fiber Surface: Preparation, Characterization, and Selective Adsorption of Li^+, Na^+, and K^+ from Aqueous Solution. Journal of Colloid And Interface Science, 246(2), 343–347.

Borho, M., Wilderer, P. (1997). Reliable method for preservation and determination of arsenate (III) concentrations in groundwater and water works. Journal of water supply rersearch and technology-Aqua, 46(3), 138–43.

Bowell, R.J. (1994). Sorption of arsenic by iron oxides and oxyhydroxides in soils. Applied geochemistry, 9(3), 279–286.

Brandriss, M.E., O'Neil, J.R., Edlund, M.B., Stoermer, E.F. (1998). Oxygen isotope fractionation between diatomaceous silica and water. Geochimica et Cosmochimica Acta, 62(7), 1119–1125.

Button, M., Moriarty, M.M., Watts, M.J., Zhang, J., Koch, I., Reimer, K.J. (2011). Arsenic speciation in field-collected and laboratory-exposed earthworms Lumbricus terrestris. Chemosphere, 85(8), 1277–1283.

Carothers, W.W., Adami, L.H., Rosenbauer, R.J. (1988). Experimental oxygen isotope fractionation between siderite-water and phosphoric acid liberated CO2-siderite. Geochimica et Cosmochimica Acta, 52(10), 2445–2450.

Casentini, B., Hug, S.J., Nikolaidis, N.P. (2011). Arsenic accumulation in irrigated agricultural soils in Northern Greece. The Science of the total environment, 409(22), 4802–4810.

Caumette, G., Koch, I., Estrada, E., Reimer, K.J. (2011). Arsenic Speciation in Plankton Organisms from Contaminated Lakes: Transformations at the Base of the Freshwater Food Chain. Environmental Science & Technology, 45(23), 9917–9923.

Chabani, M., Amrane, A., Bensmaili, A. (2007). Kinetics of nitrates adsorption on Amberlite IRA 400 resin. Desalination, 206(1-3), 560–567.

Chanda, M., O'driscoll, K.F., Rempel, G.L. (1988a). Ligand exchange sorption of arsenate and arsenite anions by chelating resins in ferric ion form I. Weak-base chelating resin dow XFS-4195. Reactive Polymers, Ion Exchangers, Sorbents, 7(2), 251–261.

Chanda, M., O'driscoll, K.F., Rempel, G.L. (1988b). Ligand exchange sorption of arsenate and arsenite anions by chelating resins in ferric ion form: II. Iminodiacetic chelating resin Chelex 100. Reactive Polymers, Ion Exchangers, Sorbents, 8(1), 85–95.

Charlet, L., Polya, D.A. (2006). Arsenic in shallow, reducing groundwaters in southern Asia: an environmental health disaster. Elements. Elements, 2(2), 91–96.

Charlet, L., Chakraborty, S., Appelo, C.A.J., Roman-Ross, G., Nath, B., Ansari, A.A., et al. (2007). Chemodynamics of an arsenic "hotspot" in a West Bengal aquifer: A field and reactive transport modeling study. Applied geochemistry, 22(7), 1273–1292.

Chiba, H., Sakai, H. (1985). Oxygen isotope exchange rate between dissolved sulfate and water at hydrothermal temperatures. Geochimica et Cosmochimica Acta, 49(4), 993–1000.

Chung, C.J., Huang, Y.L., Huang, Y.K., Wu, M.M., Chen, S.Y., Hsueh, Y.M., Chen, C.J. (2013). Urinary arsenic profiles and the risks of cancer mortality A population-based 20-year follow-up study in arseniasis-endemic areas in Taiwan. Environmental Research, 122, 25–30.

Clayton, R.N. (1961). Oxygen Isotope Fractionation between Calcium Carbonate and Water. The Journal of Chemical Physics, 34(3), 724–726.

Clayton, R.N., Mayeda, T.K. (1963). The use of bromine pentafluoride in the extraction of oxygen from oxides and silicates for isotopic analysis. Geochimica et Cosmochimica Acta, 27(1), 43–52.

Clesceri, L. S., Greensburg, A. E., Eaton, A. D.: 1998, Standard Methods for the Examination of Water and Wastewater, United Book Press, Inc., Baltimore, MD, pp. 9-134–9-135.

Cole, D.R. (2000). Isotopic exchange in mineral-fluid systems. IV. The crystal chemical controls on oxygen isotope exchange rates in carbonate-H_2O and layer silicate-H_2O systems. Geochimica et Cosmochimica Acta, 64(5), 921–931.

Cole, D.R., & Chakraborty, S. (2001). Rates and Mechanisms of Isotopic Exchange. Reviews in Mineralogy and Geochemistry, 43(1), 83–223.

Colman, A.S., Blake, R.E., Anna, M. (2002). ™^{18}O of PO_4 as a hydrogeologic tracer for nutrient contamination and subsurface microbial acticity. Denver Annual Meeting, No. 69-12.

Colman, A. S., Blake, R. E., Karl, D. M., Fogel, M. L., & Turekian, K. K. (2005). Marine phosphate oxygen isotopes and organic matter remineralization in the oceans. Proc Natl Acad Sci USA, 102(37), 13023–13028.

Consumer reports. (2012). Arsenic in your juice. How much is too much? Federal limits don't exist. Consumer Reports Magazine; Consumers Union of U.S., Inc.: Yonkers, NY, 2012. http://www.consumerreports.org/cro/magazine/2012/01/arsenic-in-your-juice/index.htm

Coplen, T.B., Böhlke, J.K., De Bievre, P., Ding, T. (2002). Isotope-abundance variations of selected elements (IUPAC Technical Report). Pure and Applied Chemisry, 74(10), 1987–2017.

Coplen, T.B., Kendall, C., Hopple, J. (1983). Comparison of stable isotope reference samples. Nature, 302(5905), 236–238.

Cornilsen, B.C., Reed, J.S. (1979). Homogeneous precipitation of basic aluminium salts as precursors for alumina. American Ceramic Society Bulletin, 58(12), 1199.

Cotton, F.A., Wilkinson, G. (1962). "Advanced inorganic chemistry: a comprehensive text", Interscience, New York., A.E. Martell, R.M. Smith (Eds.), Critical Stability Constant, Vol. 4, Prenum Press, New York and London, 1979, p. 133.

Crowson, R.A., Showers, W.J., Wright, E.K., Hoering, T.C. (1991). Preparation of phosphate samples for oxygen isotope analysis. Analytical Chemistry, 63(20), 2397–2400.

Cubadda, F., Aureli, F., D'Amato, M. (2012). Speciated urinary arsenic as a biomarker of dietary exposure to inorganic arsenic in residents living in high-arsenic areas in Latium, Italy. Pure Applied Chemistry, 84(2), 203–214.

Cui, H., Li, Q., Gao, S., Shang, J.K. (2012). Strong adsorption of arsenic species by amorphous zirconium oxide nanoparticles. Journal of Industrial and Engineering Chemistry, 18(4), 1418–1427.

Dambies, L., Salinaro, R., Alexandratos, S.D. (2004). Immobilized N-Methyl- d-glucamine as an Arsenate-Selective Resin. Environmental Science & Technology, 38(22), 6139–6146.

Dettman, D.L., Kohn, M.J., Quade, J., Ryerson, F.J., Ojha, T.P., Hamidullah, S. (2001). Seasonal stable isotope evidence for a strong Asian monsoon throughout the past 10.7 m.y. Geology, 29(1), 31.

Dietzel, M., Tang, J., Leis, A., Köhler, S.J. (2009). Oxygen isotopic fractionation during inorganic calcite precipitation - Effects of temperature, precipitation rate and pH. Chemical Geology, 268(1-2), 107–115.

BGS and DPHE. 2001. Arsenic contamination of groundwater in Bangladesh. Kinniburgh, D G and Smedley, P L (Editors). British Geological Survey Technical Report WC/00/19. British Geological Survey: Keyworth.

Eiche, E. (2002). Arsenic mobilization processes in the red river delta, Vietnam. Dissertation.

Elizalde-González, M.P., Mattusch, J., Einicke, W. D., Wennrich, R. (2001). Sorption on natural solids for arsenic removal. Chemical Engineering Journal, 81(1-3), 187–195.

Emsley, J. (2011). Nature's building blocks: an AZ guide to the elements. s. Oxford: Oxford University Press. pp. 43.

EPA, U.S. (2001). National Primary Drinking Water Regulations; Arsenic and Clarifications to Compliance and New Source Contaminants Monitoring: Final Rule, 40 CFR Parts 9, 141, and 142.

Ezeh, V.C., Harrop, T.C. (2013). Synthesis and Properties of Arsenic(III)-Reactive Coumarin-Appended Benzothiazolines: A New Approach for Inorganic Arsenic Detection. Inorganic Chemistry, 52(5), 2323–2334.

Firsching, F.H. (1961). Precipitation of silver phosphate from homogenous solution. Analytical Chemistry, 33(7), 873–874.

Freundlich, H. (1906). Over the adsorption in solution. J Phys Chem., 57A: 385-470.

Garcia-Manyes, S. (2002). Arsenic speciation in contaminated soils. Talanta, 58(1), 97–109.

Garrels, R.M., & Christ, C.L. (1965). Minerals, solutions, and equilibria. Harperand Row, 1st edition, New York, pp.450.

Goeckerman, W.H. (1940). Arsenci as the cause of cancer of mocous membran-erport of a case. Archives of Dermatology, 42(4), 641.

Gokcen, N.A. (1989). The As (Arsenic) system. Bulletin of Alloy Phase Diagrams, 10(1), 11–22.

Gouveia, M.A., Prudencio, M.I., Barros, J.S., Morgado, L., Cabral, J.M.P. (1994). Elemental concentration data for USGS Geochemical Exploration Reference Materials

GXR-1 to GXR-4 and GXR-6. Journal of Radioanalytical and Nuclear Chemistry, 179(1), 165–172.

Gómez-Ariza, J.L., Sánchez-Rodas, D., Giráldez, I., Morales, E. (2000). A comparison between ICP-MS and AFS detection for arsenic speciation in environmental samples. Talanta, 51(2), 257–268.

Gresser, M.J., Tracey, A.S., Parkinson, K.M. (1986). Vanadium (V) oxyanions: The interaction of vanadate with pyrophosphate, phosphate, and arsenate. Journal of the American Chemical Society, 108(20), 6229–6234.

Guo, H., Stüben, D., Berner, Z. (2007a). Adsorption of arsenic(III) and arsenic(V) from groundwater using natural siderite as the adsorbent. Journal of Colloid And Interface Science, 315(1), 47–53.

Guo, H., Stüben, D., Berner, Z. (2007b). Removal of arsenic from aqueous solution by natural siderite and hematite. Applied geochemistry, 22 (5), 1039-1051.

Guo, H., Stüben, D., Berner, Z., Kramar, U. (2008a). Adsorption of arsenic species from water using activated siderite-hematite column filters. Journal of Hazardous Materials, 151(2-3), 628–635.

Guo, H., Yang, S., Tang, X., Li, Y., Shen, Z. (2008b). Groundwater geochemistry and its implications for arsenic mobilization in shallow aquifers of the Hetao Basin, Inner Mongolia. The Science of the total environment, 393(1), 131–144.

Guo, X., Chen, F. (2005). Removal of arsenic by bead cellulose loaded with iron oxyhydroxide from groundwater. Environmental Science & Technology, 39(17), 6808–6818.

Halim, M.A., Majumder, R.K., Nessa, S.A., Hiroshiro, Y., Uddin, M.J., Shimada, J., Jinno, K. (2009). Hydrogeochemistry and arsenic contamination of groundwater in the Ganges Delta Plain, Bangladesh. Journal of Hazardous Materials, 164(2-3), 1335–1345.

Haran, M.J., Wasay, S.A., Tokunaga, S. (1997). Preparation of basic yttrium carbonate for phosphate removal. Water environment research, 69(5), 1047–1051.

Harland, C.E. (1994). Ion exchange: theory and practice.

Han Z.Y., Cai W., Zhang F. (2007). The current foreign research situation of high arsenic groundwater and suggestion for investigation of high arsenic groundwater in China[J]. Hydrogeology & Engineering Geology, 23, 126-128 (in Chinese).

Harvey, C.F., Swartz, C.H., Badruzzaman, A.B.M., Keon-Blute, N., Yu, W., Ali, M.A., et al. (2002). Arsenic mobility and groundwater extraction in Bangladesh. Science (New York, N.Y.), 298(5598), 1602–1606.

Hassler, J.W. (1951). Hassler: Activated carbon - Google Scholar. Chmeical Publishing Company.

He, B., Yun, Z., Shi, J., Jiang, G. (2012). Research progress of heavy metal pollution in China: Sources, analytical methods, status, and toxicity. Chinese Science Bulletin, 58(2), 134–140.

Heinrichs, H., Herrmann, A.G. (1998): Praktikum der Analytischen Geochemie. 1. Aufl. Berlin: Springer.

Hesperides. (2007). The Chemical Age - Chemical Dictionary - Chemical Terms.

Hlavay, J., Polyák, K. (2005). Determination of surface properties of iron hydroxide-coated alumina adsorbent prepared for removal of arsenic from drinking water. Journal of Colloid And Interface Science, 284(1), 71–77.

Hoefs, J. (2009). Stable isotope geochemistry. Springer.

Hollemann, A.F. Wiberg, E. (1985). Lehrbuch der Anorganischen Chemie, 91–100. Auflage. de Gruyter, Berlin, p. 919.

Holm, T.R. (2002). Effects of CO_3^{2-}/bicarbonate, Si, and PO_4^{3-} on arsenic sorption to HFO. Journal-American Water Works Association, 94(4), 174–181.

Horita, J., Clayton, R.N. (2007). Comment on the studies of oxygen isotope fractionation between calcium carbonates and water at low temperatures by Zhou and Zheng (2003; 2005). Geochimica et Cosmochimica Acta, 71(12), 3131–3135.

Hughes, M.F., Beck, B.D., Chen, Y., Lewis, A.S., Thomas, D.J. (2011). Arsenic Exposure and Toxicology: A Historical Perspective. Toxicological Sciences, 123(2), 305–332.

IBC Advanced Technologies, 2006. AnaLig _ Data Sheet: As-01 PA. IBC Advanced Technologies, Inc., Utah, USA.

IBC Advanced Technologies, 2007. AnaLig _ Data Sheet: TE-01 and TE-02. IBC Advanced Technologies, Inc., Utah, USA.

IBC Advanced Technologies, 2009. AnaLig _ Data Sheet: AN-01 Si. IBC Advanced Technologies, Inc., Utah, USA.

Jackson, B.P., Taylor, V.F., Karagas, M.R., Punshon, T., Cottingham, K. L. (2012). Arsenic, Organic Foods, and Brown Rice Syrup. Environmental Health Perspectives, 120(5), 623–626.

Jander, G., Blasius, E. (1988): Lehrbuch der analytischen und präparativen anorganischen Chemie. S. Hirzel Verlag, Stuutgart, 12 Auflage, pp.339.

Jain, A., Loeppert, R.H. (2000). Effect of competing anions on the adsorption of arsenate and arsenite by ferrihydrite. Journal of environmental quality. 29(5), 1422-1430.

Jaisi, D.P., Blake, R.E., Kukkadapu, R.K. (2010). Fractionation of oxygen isotopes in phosphate during its interactions with iron oxides. Geochimica et Cosmochimica Acta, 74(4), 1309–1319.

Jang, M., Hwang, J.S., Choi, S.I. (2007). Sequential soil washing techniques using hydrochloric acid and sodium hydroxide for remediating arsenic-contaminated soils in abandoned iron-ore mines. Chemosphere, 66(1), 8–17.

Jin, Y., Liang, C., He, G., Cao, J. (2002). Study on distribution of endemic arsenism in China [J]. Journal of Hygiene Research, 17: 517-568 (in Chinese with English abstract).

Jitmanee, K., Oshima, M., Motomizu, S. (2005). Speciation of arsenic(III) and arsenic(V) by inductively coupled plasma-atomic emission spectrometry coupled with preconcentration system. Talanta, 66(3), 529–533.

Johnson, M.O., Cohly, H., Isokpehi, R.D. (2010). The case for visual analytics of arsenic concentrations in foods. International Journal of Environmental Research and Public Health, 7, 1970-1983.

Kabata-Pendias, A. (2000). Trace elements in soils and plants. Third Edition, CRC Press.

Kaneko, M., Poulson, S.R. (2013). The rate of oxygen isotope exchange between nitrate and water. Geochimica et Cosmochimica Acta, 118, 148–156.

Kanel, S.R., Manning, B., Charlet, L., Choi, H. (2005). Removal of arsenic (III) from groundwater by nanoscale zero-valent iron. Environmental Science & Technology, 39(5), 1291–1298.

Karthikeyan, S., Hirata, S. (2003). Simultaneous determination of arsenic(III) and arsenic(V) by flow injection–inductively coupled plasma–atomic emission spectrometry (ICP-AES) with ultrasonic nebulization. Analytical and Bioanalytical Chemistry, 375(1), 139–144.

Khalid, N., Ahmad, S., Toheed, A.,Ahmed, J. (1998). Immobilization of arsenic on rice husk. Adsorption science & technology, 16(8), 655–666.

Kim, Y.T., Yoon, H.O., Yoon, C., Woo, N.C. (2009). Arsenic species in ecosystems affected by arsenic-rich spring water near an abandoned mine in Korea. Environmental pollution (Barking, Essex : 1987), 157(12), 3495–3501.

Kolodny, Y., Luz, B., Navon, O. (1983). Oxygen isotope variations in phosphate of biogenic apatites, I. Fish bone apatite—rechecking the rules of the game. Earth and Planetary Science Letters, 64(3), 398–404.

Kornexl, B., Gehre, M., Hofling, R., Werner, R. (1999a). On-line delta(18)O measurement of organic and inorganic substances. Rapid Communications in Mass Spectrometry, 13(16), 1685–1693.

Kornexl, B., Werner, R., Gehre, M. (1999b). Standardization for oxygen isotope ratio measurement - still an unsolved problem. Rapid Communications in Mass Spectrometry, 13(13), 1248–1251.

Koziet, J. (1997). Isotope ratio mass spectrometric method for the on-line determination of oxygen-18 in organic matter. Journal of Mass Spectrometry, 32(1), 103–108.

Kroopnick, P., & Craig, H. (1972). Atmospheric Oxygen: Isotopic Composition and Solubility Fractionation. Science (New York, N.Y.), 175(4017), 54–55.

Ladeira, A.C.Q., Ciminelli, V. S. T. (2004). Adsorption and desorption of arsenic on an oxisol and its constituents. Water Research, 38(8), 2087–2094.

Langmuir, I. (1917). The constitution and fundamental properties of solids and liquids. Journal of the American Chemical Society, 39(9), 1848–1906.

LaPorte, D.F., Holmden, C., Patterson, W.P., Prokopiuk, T., Eglington, B.M. (2009). Oxygen isotope analysis of phosphate: improved precision using TC/EA CF‐IRMS. Journal of Mass Spectrometry, 44(6), 879–890.

Larese-Casanova, P., Blake, R.E. (2012). Measurement of $\delta^{18}O$ values in arsenic and selenium oxyanions. Rapid Communications in Mass Spectrometry, 27(1), 117–126.

Lasaga, A.C. (1998). Kinetic theory in the earth sciences. Princeton series in geochemistry, ISSN 1931-0935.

Lecuyer, C., Fourel, F., Martineau, F., Amiot, R., Bernard, A., Daux, V., et al. (2007). High-precision determination of $^{18}O/^{16}O$ ratios of silver phosphate by EA-pyrolysis-IRMS continuous flow technique. Journal of Mass Spectrometry, 42(1), 36–41.

Lecuyer, C., Grandjean, P., Sheppard, S.M. (1999). Oxygen isotope exchange between dissolved phosphate and water at temperatures≤ 135 °C: Inorganic versus biological fractionations. Geochimica et Cosmochimica Acta, 63(6), 855–862.

Lecuyer, C., Reynard, B., Grandjean, P. (2004). Rare earth element evolution of Phanerozoic seawater recorded in biogenic apatites. Chemical Geology, 204(1-2), 63–102.

Lee, S.O., Tran, T., Jung, B.H., Kim, S.J., Kim, M.J. (2007). Dissolution of iron oxide using oxalic acid. Hydrometallurgy, 87(3-4), 91–99.

Lenoble, V., Laclautre, C., Deluchat, V., Serpaud, B., Bollinger, J.C. (2005). Arsenic removal by adsorption on iron(III) phosphate. Journal of Hazardous Materials, 123(1-3), 262–268.

Lenoble, V., Laclautre, C., Serpaud, B., Deluchat, V., Bollinger, J.C. (2004). As(V) retention and As(III) simultaneous oxidation and removal on a MnO_2-loaded polystyrene resin. Science of The Total Environment, 326(1-3), 197–207.

Lester, G.W., Kyser, T.K., Clark, A.H. (2013). Oxygen isotope partitioning between immiscible silicate melts with H_2O, P and S. Geochimica et Cosmochimica Acta, 109, 306–311.

Liang, Y., Blake, R.E. (2005). Phosphate oxygen isotope ratio proxy for specific microbial activity in marine sediments (Peru Margin). AGU Fall Meeting Abstracts, -1, 0672.

Liang, Y., Blake, R.E. (2007). Oxygen isotope fractionation between apatite and aqueous-phase phosphate: 20–45 °C. Chemical Geology, 238(1-2), 121–133.

Linge, K.L. (2002). Assessment of contaminant availability in a shallow wetland. Ph.D. Dissertation, Unicersity of Western Australia.

Liu, X., Zhang, W., Hu, Y., Cheng, H. (2013). Extraction and detection of organoarsenic feed additives and common arsenic species in environmental matrices by HPLC–ICP-MS. Microchemical Journal, 108(C), 38–45.

Longinelli, A. (1984). Oxygen isotopes in mammal bone phosphate: A new tool for paleohydrological and paleoclimatological research. Geochimica et Cosmochimica Acta, 48(2), 385–390.

Lovley, D.R. (1987). Organic matter mineralization with the reduction of ferric iron: A review. Geomicrobiology Journal, 5(3-4), 375–399.

Lovley, D.R. (1993). Dissimilatory metal reduction. Annual Reviews in Microbiology, 47(1), 263–290.

Mandal, B.K., Suzuki, K.T. (2002). Arsenic round the world: a review. Talanta, 58(1), 201–235.

Mangalo, M., Meckenstock, R.U., Stichler, W., Einsiedl, F. (2007). Stable isotope fractionation during bacterial sulfate reduction is controlled by reoxidation of intermediates. Geochimica et Cosmochimica Acta, 71(17), 4161–4171.

Maréchal, C., Albarède, F. (2002). Ion-exchange fractionation of copper and zinc isotopes. Geochimica et Cosmochimica Acta, 66(9), 1499–1509.

Martinez, V.D., Vucic, E. A., Becker-Santos, D.D., Gil, L., Lam, W.L. (2011). Arsenic Exposure and the Induction of Human Cancers. Journal of Toxicology, 5576, 1–13.

Matsunaga, H., Yokoyama, T., Eldridge, R.J., Bolto, B. A. (1996). Adsorption characteristics of arsenic(III) and arsenic(V) on iron(III)-loaded chelating resin having lysine-Nα,Nα-diacetic acid moiety. Reactive and Functional Polymers, 29(3), 167–174.

McArthur, J.M., Banerjee, D.M., Hudson-Edwards, K.A., Mishra, R., Purohit, R., Ravenscroft, P., et al. (2004). Natural organic matter in sedimentary basins and its relation to arsenic in anoxic ground water: the example of West Bengal and its worldwide implications. Applied geochemistry, 19(8), 1255–1293.

McArthur, J.M., Ravenscroft, P., Safiulla, S., Thirlwall, M.F. (2001). Arsenic in groundwater: Testing pollution mechanisms for sedimentary aquifers in Bangladesh. Water Resources Research, 37(1), 109–117.

McCrea, J. M. (1950). On the isotopic chemistry of carbonates and a paleotemperature scale. The Journal of Chemical Physics, 18. 849-857.

McLaughlin, K., Silva, S., Kendall, C., Stuart-Williams, H., & Paytan, A. (2004). A precise method for the analysis of δ18O of dissolved inorganic phosphate in seawater. Limnol. Oceanogr.: Methods, 2, 202–212.

Melamed, D. (2005). Monitoring arsenic in the environment: a review of science and technologies with the potential for field measurements. Analytica Chimica Acta, 532(1), 1–13.

Mendham, J., Mendham, J. (2006). Vogels textbook of quantitative chemical analysis. Pearson Education India.

Meng, X., Wang, W. (1998). Speciation of arsenic by disposable cartridges. Third International Conference on Arsenic Exposure and Health Effects, San Diego, CA, July 12-15, 1998

Meng, X., Korfiatis, G.P., Christodoulatos, C., Bang, S. (2001a). Treatment of arsenic in Bangladesh well water using a household co-precipitation and filtration system. Water Research, 35(12), 2805–2810.

Meng, X., Korfiatis, G.P., Jing, C., Christodoulatos, C. (2001b). Redox Transformations of Arsenic and Iron in Water Treatment Sludge during Aging and TCLP Extraction, 35(17), 3476–3481.

Mermet, J. M. (2005). Is it still possible, necessary and beneficial to perform research in ICP-atomic emission spectrometry? Journal of Analytical Atomic Spectrometry 20, 11–16.

Mizutani, Y., Rafter, T.A. (1969). Oxygen isotopic composition of sulphates—Part 3: Oxygen isotopic fractionation in the bisulphate ion–. New Zealand Journal of Science 12, 54–59.

Mohan, D., Pittman, C.U., Jr. (2007). Arsenic removal from water/wastewater using adsorbents - A critical review. Journal of Hazardous Materials, 142(1-2), 1–53.

Mohr, P.J., Taylor, B.N., Newell, D.B. (2008). CODATA recommended values of the fundamental physical constants: 2006. Reviews of Modern Physics, 80(2), 633–730.

Mondal, P., Majumder, C. B., Mohanty, B. (2006). Laboratory based approaches for arsenic remediation from contaminated water: recent developments. Journal of Hazardous Materials, 137(1), 464–479.

Moreira, C.M., Duarte, F.A., Lebherz, J., Pozebon, D., Flores, E.M.M., Dressler, V.L. (2011). Arsenic speciation in white wine by LC–ICP–MS. Food chemistry, 126(3), 1406–1411.

Mukherjee, A., Bhattacharya, P., Shi, F., Fryar, A.E., Mukherjee, A.B., Xie, Z.M., et al. (2009). Chemical evolution in the high arsenic groundwater of the Huhhot basin (Inner Mongolia, PR China) and its difference from the western Bengal basin (India). Applied geochemistry, 24(10), 1835–1851.

Muñoz, J.A., Gonzalo, A., & Valiente, M. (2002). Arsenic Adsorption by Fe(III)-Loaded Open-Celled Cellulose Sponge. Thermodynamic and Selectivity Aspects. Environmental Science & Technology, 36(15), 3405–3411.

Mustafa, S., Ahmad, T., Naeem, A., Shah, K.H., Waseem, M. (2009). Kinetics of Chromium Ion Removal from Tannery Wastes Using Amberlite IRA-400 Cl− and its Hybrids. Water, Air, & Soil Pollution, 210(1-4), 43–50.

Mustafa, S., Hussain, S.Y., Ahmad, R. (1990). Phosphate/hydroxide exchange studies on Amberlite IRA-400. Solvent Extraction and Ion Exchange, 8(2), 325–340.

Mustafa, S., Hussain, S.Y., & Khan, A.N. (1992). Kinetics of ion exchange sorption of phosphate. Reaction Kinetics & Catalysis Letters, 47(1), 125–131.

Mustafa, S., Hussain, S.Y., Rehana, N., Alamzeb. (1989). Temperature effect on ion exchange sorption of phosphate. Solvent Extraction and Ion Exchange, 7(4), 705–720.

Mustafa, S., Naeem, A., Rehana, N., Hamid, A., Dilara, B. (2004). Phosphate/Sulphate Exchange Studies On Amberlite IRA-400. Environmental Technology, 25(10), 1115–1122.

Nam, S.H., Oh, H.J., Min, H.S., Lee, J.H. (2010). A study on the extraction and quantitation of total arsenic and arsenic species in seafood by HPLC–ICP-MS. Microchemical Journal, 95(1), 20–24.

Namasivayam, C., Senthilkumar, S. (1998). Removal of Arsenic(V) from Aqueous Solution Using Industrial Solid Waste: Adsorption Rates and Equilibrium Studies. Industrial & engineering chemistry research, 37(12), 4816–4822.

Narcise, C., Coo, L., Delmundo, F. (2005). On-line preconcentration and speciation of arsenic by flow injection hydride generation atomic absorption spectrophotometry. Talanta, 68(2), 298–304.

Ng, J.C., Wang, J., Shraim, A. (2003). A global health problem caused by arsenic from natural sources. Chemosphere, 52(9), 1353–1359.

Nickson, R., McArthur, J., Burgess, W., Ahmed, K.M., Ravenscroft, P., Rahman, M. (1998). Arsenic poisoning of Bangladesh groundwater. Nature, 395(6700), 338.

Niethardt, H. (2012). Impact of groundwater abstraction and of the availability of organic matter on the release and distribution of arsenic in groundwater of the Bengal Delta Plain, India.

Nalcaci, O.O., Ruzin, E., Bockhorn, H., 2009, Synthesis of nano-sized iron oxide particles in low-pressure hydrogen flames, European Aerosol Conference Proceedings, Karlsruhe, Germany, 1, 48.

Nalcaci, O.O., Akten, D., Cinar, D., Bockhorn, H., 2010a, A flame model for the investigation of oxidic nanoparticles, X International Conference on Nanostructured Materials Proceedings, Rome, Italy, 27.

Nalcaci, O.O., Akten, D., Bockhorn, H., (2010b). Effect of precursor concentration on the formation of titania particles in a low pressure hydrogen flame, X International Conference on Nanostructured Materials Proceedings, Rome, Italy, 91.

Norman, N.C. (1998). Chemistry of arsenic, antimony and bismuth. Springer. p. 50.

Norra, S., Berner, Z.A., Agarwala, P., & Wagner, F. (2005). Impact of irrigation with As rich groundwater on soil and crops: A geochemical case study in West Bengal Delta Plain, India. Applied Geochemistry, 20(10), 1890–1906.

Northrop, D.A., Clayton, R.N. (1966). Northrop and Clayton 1966-oxygen isotope fractionationas in systems containing dolomite. The Journal of Geology, 174–196.

Ohmoto, H., Lasaga, A.C. (1982). Kinetics of reactions between aqueous sulfates and sulfides in hydrothermal systems. Geochimica et Cosmochimica Acta, 46(10), 1727–1745.

Okumura, A., Okazaki, N. (1973). The kinetics of oxygen exchange between arsenate ions and solvent water. Bull. Chem. Soc. Jpn, 46, 2937-2941.

Okumura, A., Matsumiya, Y., Yamamoto, K., Ueno, R., Suzuki, M., Yamabe, S. (1995). Kinetics of oxygen exchange between arsenic acid and solvent water. Bulletin of the chemical society of japan, 68(7), 1839–1849.

Onishi, H. (1969). Arsenic, in: K.H. Wedepohl (Ed.), Handbook of Geochemistry, Springer-Verlag, New York, 1969, Vol. II-2, Chapter 33.

Oremland, R.S., Stolz, J.F. (2005). Arsenic, microbes and contaminated aquifers. Trends in microbiology, 13(2), 45–49.

Oura, K., Saranin, A.A., Zotof, A.V. (2003). Surface science: an introduction. Berlin: Springer.

O'Neil, J.R., Roe, L.J., Reinhard, E., Blake, R. E. (1994). A rapid and precise method of oxygen isotope analysis of biogenic phosphate. Israel Journal of Earth Sciences, 43(3-4), 203–212.

O'Neil, J.R., Vennemann, T.W., McKenzie, W.F. (2003). Effects of speciation on equilibrium fractionations and rates of oxygen isotope exchange between $(PO_4)_{aq}$ and H_2O. Geochimica et Cosmochimica Acta, 67(17), 3135–3144.

Packer, A.P., Ciminelli, V.S. (2005). A simplified flow system for inorganic arsenic speciation by preconcentration of As(V) and separation of As(III) in natural waters by ICP-MS. Atomic Spectroscopy, 26(4), 131.

Pakshirajan, K., Associate, Izquierdo, M., Lens, P.N.L. (2013). Arsenic(III) Removal at Low Concentrations by Biosorption using Phanerochaete chrysosporiumPellets. Separation Science and Technology, 48(7), 1111–1122.

Parkhurst, D.L., Appelo, C. A.J. (1999). User's Guide to PHREEQC (version 2)- User's guide to PHREEQC (version 2)-A computer program for speciation, batch-reaction, one-dimensional transport, and inverse geochemical calculations: U.S. Geological Survey Water-Resources Investigations Report 99-4259, 312 p.

Pavicevic, M.K., Amthauer, G. (2000): Physikalisch-chemische Untersuchungsmethoden in den Geowissenschaften, Band 1: Mikroskopische, analytische und massenspektrometrische Methoden. Stuttgart : Schweizerbart.

Pierce, M.L., & Moore, C.B. (1982). Adsorption of arsenite and arsenate on amorphous iron hydroxide. Water Research, 16(7), 1247–1253.

Pizarro, I., Gómez, M., Cámara, C., Palacios, M.A. (2003). Arsenic speciation in environmental and biological samples. Analytica Chimica Acta, 495(1-2), 85–98.

Rahman, I.M.M., Begum, Z.A., Nakano, M., Furusho, Y., Maki, T., Hasegawa, H. (2011). Selective separation of arsenic species from aqueous solutions with immobilized macrocyclic material containing solid phase extraction columns. Chemosphere, 82(4), 549–556.

Rahmani, A., Ghaffari, H., Samadi, M. (2011). A Comparative study on arsenic (III) removal from aqueous solution using nano and micro sized zero-valent iron. Iranian Journal of Environmental Health Science & Engineering, 8(2).

Rakestraw, N.M., Rudd, D.P., Dole, M. (1951). Isotopic composition of oxygen in air dissolved in Pacific Ocean water as a function of depth. Journal of the American Chemical Society, 73(6), 2976–2976.

Ravenscroft, P., McArthur, J.M., Hoque, B.A. (2001). Geochemical and palaeohydrological controls on pollution of groundwater by arsenic.

Reddy, K.J., McDonald, K.J., King, H. (2013). A novel arsenic removal process for water using cupric oxide nanoparticles. Journal of Colloid And Interface Science, 397(C), 96–102.

Richmond, W.R., Loan, M., Morton, J., Parkinson, G. M. (2004). Arsenic removal from aqueous solution via ferrihydrite crystallization control. Environmental Science & Technology, 38(8), 2368–2372.

Rivas, B.L., Muñoz, C. (2009). Functional water-insoluble polymers with ability to remove arsenic(V). Polymer Bulletin, 65(1), 1–11.

Rosman, K.J.R., Taylor, P.D.P. (1998). Isotopic Compositions of the Elements 1997. Journal of Physical and Chemical Reference Data, 27(6), 1275–1287.

Ross, H.H., Hahn, R.B., (1961) A study of the separation of phosphate ion from arsenate ion by sllvent extraction. Taianta, 7: 276 - 280.

Sahoo, P.K., Kim, K. (2013). A review of the arsenic concentration in paddy rice from the perspective of geoscience. Geosciences Journal, 17(1), 107–122.

Santrock, J., Hayes, J.M. (1987). Adaptation of the Unterzaucher procedure for determination of oxygen-18 in organic substances. Analytical Chemistry, 59(1), 119–127.

Schwander, H. (1953). Bestimmung des relativen Sauerstoffisotopen-Verha¨ltnisses in Silikatgesteinen und -Mineralien. Geochimica et Cosmochimica Acta, 4, 261–291.

Sharma, V.K., Sohn, M. (2009). Aquatic arsenic: Toxicity, speciation, transformations, and remediation. Environment International, 35(4), 743–759.

Secretaria de Salud y Asistencia (SSA), (1999). Modificación a la norma official mexicana NOM-127-SSA1-1994, salud ambiental. Agua para uso y consume humano. Límites permisibles de calidad y tratamientos a que debe someterse el agua para su potabilización, Mexico D.F. <www.salud.gob.mx/unidades/cdi/nom/m127ssa14.html>.

Siegfried, K., Endes, C., Bhuiyan, A.F.M.K., Kuppardt, A., Mattusch, J., van der Meer, J.R., et al. (2012). Field Testing of Arsenic in Groundwater Samples of Bangladesh Using a Test Kit Based on Lyophilized Bioreporter Bacteria. Environmental Science & Technology, 46(6), 3281–3287.

Silver, S., Phung, L.T. (2005). A bacterial view of the periodic table: genes and proteins for toxic inorganic ions. Journal of industrial microbiology & biotechnology, 32(11-12), 587–605.

Singleton, M.J., Woods, K.N., Conrad, M.E., Depaolo, D.J., Dresel, P.E. (2005). Tracking sources of unsaturated zone and groundwater nitrate contamination using nitrogen and oxygen stable isotopes at the Hanford site, Washington. Environmental Science & Technology, 39(10), 3563–3570.

Smedley, P.L., Kinniburgh, D.G. (2002). A review of the source, behaviour and distribution of arsenic in natural waters. Applied geochemistry, 17(5), 517–568.

Smith, A.H., Lingas, E.O., Rahman, M. (2000). Contamination of drinking-water by arsenic in Bangladesh: a public health emergency. Bulletin of the World Health Organization, 78(9), 1093–1103.

Soner Altundoğan, H., Altundoğan, S., Tümen, F., & Bildik, M. (2000). Arsenic removal from aqueous solutions by adsorption on red mud. Waste Management, 20(8), 761–767.

Sordelet, D., Akinc, M. (1988). Preparation of spherical, monosized Y2O3 precursor particles. Journal of Colloid And Interface Science, 122(1), 47–59.

Stephan, E. (2000). Oxygen Isotope Analysis of Animal Bone Phosphate: Method Refinement, Influence of Consolidants, and Reconstruction of Palaeotemperatures for Holocene Sites. Journal of Archaeological Science, 27(6), 523–535.

Stefánsson, A., Gunnarsson, I., Giroud, N. (2007). New methods for the direct determination of dissolved inorganic, organic and total carbon in natural waters by Reagent-Free Ion Chromatography and inductively coupled plasma atomic emission spectrometry. Analytica Chimica Acta 582 (1), 69–74.

Stern, J.C., Wang, Y. (2002). Using oxygen isotopes to determine the source of phosphate in the Everglades. AGU Spring Meeting Abstracts.

Stüben, D., Berner, Z., Chandrasekharam, D., Karmakar, J. (2003). Arsenic enrichment in groundwater of West Bengal, India: geochemical evidence for mobilization of As under reducing conditions. Applied geochemistry, 18(9), 1417–1434.

Suzuki, T.M., Bomani, J.O., Matsunaga, H. (1997). Removal of As(III) and As(V) by a porous spherical resin loaded with monoclinic hyfrous zirconnium oxide. Chemistry Letters, (11), 1119-1120.

Suzuki, T.M., Bomani, J.O., Matsunaga, H., Yokoyama, T. (2000). Preparation of porous resin loaded with crystalline hydrous zirconium oxide and its application to the removal of arsenic. Reactive and Functional Polymers, 43(1), 165–172.

Suzuki T.M., Tanco, M.L., Tanaka, D.A.P., Matsunaga, H., Yokoyama, T. (2001). Adsorption characteristics and oxo-anions of arsenic and selenium on the porous polymers loaded with monoclinic hydrous zirconum oxide. Separation Science and Technology, 36(1), 103–111.

Tamburini, F., Bernasconi, S.M., Angert, A., Weiner, T., Frossard, E. (2010). A method for the analysis of the $\delta^{18}O$ of inorganic phosphate extracted from soils with HCl. European Journal of Soil Science, 61(6), 1025–1032.

Tanada, M., Miyoshi, T., Nakamura, T., Tanada, S. (1990). Adsorption removal of cresol by granular activated carbon for medical waste water treatment. Bulletin of environmental contamination and toxicology, 45(2), 170–176.

Tang, W., Li, Q., Gao, S., Shang, J.K. (2011). Arsenic (III,V) removal from aqueous solution by ultrafine α-Fe2O3 nanoparticles synthesized from solvent thermal method. Journal of Hazardous Materials, 192(1), 131–138.

Taylor, J.W., & Chen, I.J. (1970). Variables in oxygen-18 isotopic analysis by mass spectrometry. Analytical Chemistry, 42(2), 224–228.

Teng, H., Hsieh, C.T. (1998). Influence of surface characteristics on liquid-phase adsorption of phenol by activated carbons prepared from bituminous coal. Industrial & engineering chemistry research, 37(9), 3618–3624.

Thakur, J.K., Thakur, R.K., Ramanathan, A.L., Kumar, M., Singh, S.K. (2011). Arsenic Contamination of Groundwater in Nepal—An Overview. Water, 3(1), 1–20.

Toth, J. (2002). Adsorption: Theory, Modeling, and Analysis. Surfactant Science Series 107, Dekker, New York..

Treybal, R.E. (1988). Mass-transfer Operations. McGrew-Hill, New York.

Trung, D.Q., Anh, C.X., Trung, N.X., Yasaka, Y., Fujita, M., Tanaka, M. (2001). Preconcentration of arsenic species in environmental waters by solid phase extraction using metal-loaded chelating resins. Analytical sciences, 17, i1219–i1222.

Tudge, A.P. (1960). A method of analysis of oxygen isotopes in orthophosphate—its use in the measurement of paleotemperatures. Geochimica et Cosmochimica Acta, 18, 81–93.

Tuzen, M., Citak, D., Mendil, D., Soylak, M. (2009). Arsenic speciation in natural water samples by coprecipitation-hydride generation atomic absorption spectrometry combination. Talanta, 78(1), 52–56.

Usdowski, E., Hoefs, J. (1993). Oxygen isotope exchange between carbonic acid, bicarbonate, carbonate, and water: A re-examination of the data of McCrea (1950) and an expression for the overall partitioning of oxygen isotopes between the carbonate species and water. Geochimica et Cosmochimica Acta, 57(15), 3815–3818.

Van Geen, A., Rose, J., Thoral, S., Garnier, J.M., Zheng, Y., Bottero, J.Y. (2004). Decoupling of As and Fe release to Bangladesh groundwater under reducing conditions. Part II: Evidence from sediment incubations. Geochimica et Cosmochimica Acta, 68(17), 3475–3486.

Van Hulle, M., Zhang, C., Zhang, X., & Cornelis, R. (2002). Arsenic speciation in Chinese seaweeds using HPLC-ICP-MS and HPLC-ES-MS. Analyst, 127(5), 634–640.

Vennemann, T.W., Fricke, H.C., Blake, R.E., O'Neil, J.R., Colman, A. (2002). Oxygen isotope analysis of phosphates: a comparison of techniques for analysis of Ag 3 PO 4. Chemical Geology, 185(3), 321–336.

Violante, A., Pigna, M., Ricciardella, M., Gianfreda, L. (2002). Adsorption of phosphate on variable charge minerals and soils as affected by organic and inorganic ligands. In Developments in Soil Science, 28, 279–295).

Vogel A.I., A text-book of quantitative inorganic analysis including elementary instrumental analysis. Woolwich Polytechnic, England. 3rd ed., John Wiley and Sons, Inc., New York, 1961, 270-271.

Wang, P., Zhao, G., Tian, J., Su, X. (2010). High-Performance Liquid Chromatography–Inductively Coupled Plasma Mass Spectrometry Based Method for the Determination of Organic Arsenic Feed Additives and Speciation of Anionic Arsenics in Animal Feed. Journal of Agricultural and Food Chemistry, 58(9), 5263–5270.

Wasay, S.A., Haron, M.J., Uchiumi, A., & Tokunaga, S. (1996). Removal of arsenite and arsenate ions from aqueous solution by basic yttrium carbonate. Water Research, 30(5), 1143–1148.

Werner, R. (2003). The Online $^{18}O/^{16}O$ Analysis: Development and application. Isotopes in Environmental and Health Studies, 39(2), 85–104.

Whaley-Martin, K.J., KOCH, I., Reimer, K.J. (2012). Arsenic species extraction of biological marine samples (Periwinkles, Littorina littorea) from a highly contaminated site. Talanta, 88, 187–192.

Willard, H.H., Tang, N.K. (1937). Journal of the American Chemical Society, 59(7), 1190–1196.

Woolson, E.A. (1977). Fate of arsenicals in different environmental substrates. Environmental Health Perspectives, 19, 73–81.

Xiong, C., He, M., Hu, B. (2008). On-line separation and preconcentration of inorganic arsenic and selenium species in natural water samples with CTAB-modified alkyl silica microcolumn and determination by inductively coupled plasma-optical emission spectrometry. Talanta, 76(4), 772–779.

Yadanaparthi, S.K.R., Graybill, D., Wandruszka, von, R. (2009). Adsorbents for the removal of arsenic, cadmium, and lead from contaminated waters. Journal of Hazardous Materials, 171(1-3), 1–15.

Yoshizuka, K., Nishihama, S., Sato, H. (2010). Analytical survey of arsenic in geothermal waters from sites in Kyushu, Japan, and a method for removing arsenic using magnetite. Environmental geochemistry and health, 32(4), 297–302.

Yuchi, A., Ogiso, A., Muranaka, S., Niwa, T. (2003). Preconcentration of phosphate and arsenate at sub-ng ml−1 level with a chelating polymer-gel loaded with zirconium(IV). Analytica Chimica Acta, 494(1-2), 81–86.

Zazzo, A., Lecuyer, C., Mariotti, A. (2004). Experimentally-controlled carbon and oxygen isotope exchange between bioapatites and water under inorganic and microbially-mediated conditions. Geochimica et Cosmochimica Acta, 68(1), 1–12.

Zhang, G., Qu, J., Liu, H., Liu, R., Wu, R. (2007a). Preparation and evaluation of a novel Fe–Mn binary oxide adsorbent for effective arsenite removal. Water Research, 41(9), 1921–1928.

Zhang, G.S., Qu, J.H., Liu, H.J., Liu, R.P., Li, G.T. (2007b). Removal mechanism of As(III) by a novel Fe-Mn binary oxide adsorbent: oxidation and sorption. Environmental Science & Technology, 41(13), 4613–4619.

Zhang, T., Sun D.D. (2013). Removal of arsenic from water using multifunctional micro-/nano-structured MnO_2 spheres and microfiltration. Chemical Engineering Journal, 225, 271–279.

Zheng, J., Hintelmann, H. (2009). HPLC-ICP-MS for a comparative study on the extraction approaches for arsenic speciation in terrestrial plant, Ceratophyllum demersum. Journal of Radioanalytical and Nuclear Chemistry, 280(1), 171–179.

Zheng, Y.F. (1999). Oxygen isotope fractionation in carbonate and sulfate minerals. Geochemical Joural japan, 33, 109–126.

Zheng, Y.F., Zhou, G.T. (2007). Response to the Comment by J. Horita and R.N. Clayton on "The studies of oxygen isotope fractionation between calcium carbonates and water at low temperatures." Geochimica et Cosmochimica Acta, 71(12), 3136–3143.

Zhou, G.T. (2005). Effect of polymorphic transition on oxygen isotope fractionation between aragonite, calcite, and water: A low-temperature experimental study. American Mineralogist, 90(7), 1121-1130.

Zhou, G.T., Zheng, Y.F. (2001). Kinetic mechanism of oxygen isotope disequilibrium in precipitated witherite and aragonite at low temperatures- An experimental study. Geochimica et Cosmochimica Acta, 66(12), 3195–3197.

Zhu, X., Alexandratos, S.D. (2005). Affinity and Selectivity of Immobilized N-Methyl-d-glucamine for Mercury(II) Ions. Industrial & engineering chemistry research, 44(19), 7490–7495.

Zobrist, J., Dowdle, P.R., Davis, J.A., Oremland, R.S. (2000). Mobilization of Arsenite by Dissimilatory Reduction of Adsorbed Arsenate. Environmental Science & Technology, 34(22), 4747–4753.

Zuo, J.C., Tong, S.R., Yu, X.L., Wu, L.Y., Cao, C.Y., Ge, M.F., Song, W.G. (2012). Fe^{3+} and amino functioned mesoporous silica: Preparation, structural analysis and arsenic adsorption. Journal of Hazardous Materials, 235-236, 336–342.

Appendix

Appendix 1: Chemicals and equipments

Chemicals:

- 28-30 vol% NH_4OH: p.a. MERCK
- 32 vol% NaOH: p.a. Roth
- 37 vol% HCl: p.a. Roth
- 4.6 g $LaCl_3.7H_2O$: Aldrich
- 40 vol% HF: 40 % suprapure, Merck
- 65 vol% HNO_3: suprapure, VWM
- 69 vol% HNO_3: suprapure, Roth
- 96 vol% ethanol: Roth
- 96 vol% CH_3COOH: p.a. Merck
- $AgNO_3$ solid: GR Merck
- Amberlite IRA-400 (OH): Aldrich, 20-50 mesh
- Amberlite XAD-7: Rohm & Haas Co., 32-60 mesh
- Ba $(NO_3)_2$ solid: p.a. Merck
- Cartridge with Alumino-silicate based adsorbent (MAS)： Stevens
- Institute of Technology, USA
- Deionized water: 18.2 MΩ.cm Milli-Q
- K_2SO_4: p.a. Merck
- Muromac B-1 chelating resin with Na+ form: muromachi kgaku Kogyo kaisha, Ltd, Japan
- $Na_2HAsO_4 \cdot 7H_2O$: p.a. MERCK ≥99 %
- $NaAsO_2$: p.a. MERCK ≥99 %
- NaCl: p.a. MERCK
- NaCl: p.a. Merck
- $NaNO_3$ (p.a. Merck)
- NaOH standard solution (0.25M): FLUKA
- NH_4NO_3: p.a. Merck
- NMDG： Sigma-Aldrich
- Phenolphthalein: Sigma-Aldrich

- PolyDVB: Sigma-Aldrich?
- Tap water: from the tap water supply system of Karlsruhe, Germany (http://www.stadtwerke-karlsruhe.de/swka-de/PDF/Service/Infomaterial/Produkte/Trinkwasser_Qualitaet.pdf).
- Urea (NH_2CONH_2): GR Sigma
- $YCl_3.6H_2O$: Aldrich 99 % trace metal basis
- $ZrOCl_2. 8H_2O$: GR Aldrich

Materials and equipment:

- Alkalinity Test field kit test: Acid capacity to pH 8.2 and pH 4.3
- Balancer: Sartorius SE2-F
- Brunauer–Emmett–Teller (BET): BEL Japan, Inc.
- Burette: Merck
- Cellulose Acetate Filter: Sartorius stedim biotech, (0.20 µm, 0.45 µm pore size)
- Cl^- field kit test: Merck, 5-125 mg/L
- Conductivity meter: WTW 330
- Element Analyzer (EA): EuroEA3000; Euro Vector
- Empty PP REV Tube kit: supelco
- Empty PP REV Tube kit: supelco
- Energy dispersive X-ray spectroscopy (EDX): Epsilon 5, PANanalytical

- Erlenmeyer flask: DURAN, Germany
- Fluid injection atomic spectrometry (FIAS): FIAS 400; Perkin-Elmer
- GEN bag: Biomerieux
- GEN box: Biomerieux
- GEN indicator: Biomerieux
- Glove box: Biomerieux
- High resolution Resolution-Inductively Coupled Plasma- Mass Spectrometry (ICP-MS): Element 2, Thermo Fisher, Bremen, Germany
- Inductively Coupled Plasma Mass Spectrometry -Optical Emission Spectrometry (ICP-OES): Spectro Ciros CCD, Kleve, Germany
- Ion-Chromatography: (ICS 1000; Dionex).
- Isotope ratios-Mass Spectrometry (IRMS): IsoPrime, GV Instruments
- Liquid water isotope analyzer (LWIA-24d): Los Gatos Researc
- Peristaltic pump: ISMATEC, ICP
- pH indicator: Merck

- pH meter :WTW 330
- Portable Photometer: Merck
- PVC tube: length (12 cm, diameter: 1.2 cm): Roth
- Rotary evaporator： STUART model RE-300, UK
- Scanning electric Microscopy (SEM): Zeiss DSM 960
- Shaker: KS-15 CONTROL, Edmund Buehler GmbH
- Silver cupscapsules: Perkin Elmer
- SO_4^{2-} field kit test: Merck, 5-250 mg/L
- Spectrometry: Lambda 2; PerkinElmer
- Spectrometry: Lambda 2; PerkinElmer
- Thermal Conversion/Elemental Analyzer-Gas Chromatogrphy-isotope ratio mass Spectrometry (TC/EA-GC –IRMS): Finnigan
- X-ray Diffraction (XRD): Kristalloflex D500, Siemens, Germa

Appendix 2: the raw data for the Figures in the chapters

Table A2.1: The raw data for the Figure 4.10: Comparison of As concentration after adsorption between column test and shaking tests with HMZR.

Column experiment			**Shaking experiment**		
Time (min)	**Reaction condition**	**As(III) in water (μg/L)**	**Time (min)**	**Reaction condition**	**As(III) in water (μg/L)**
Initial		119	**Initial**	Shaking	101
0	Column	-	**180**	with 1.0 g MHZR at rate of 150 rpm	1.6
30	with 1.0 g MHZR at a flow rate	0.85			
80	of 0.55	0.20			
140	mL/min	0.51			
200		0.99			
900		5.79			
1200		7.19			
1800		9.79			

Table A2.2: The raw data for the Figure 4.11: Comparison of As concentration after adsorption between column test and shaking tests with BYC.

Column experiment			**Shaking experiment**		
Time (min)	**Reaction condition**	**As(III) in water (μg/L)**	**Time (min)**	**Reaction condition**	**As(III) in water (μg/L)**
Initial		99.4	**Initial**	Shaking	101
0	Column	-	**900**	with 0.25 g BYC at rate of 150 rpm	8.7
30	with 0.25	28.4			
60	g BYC at a flow rate	31.6			
20	of 0.85	38.1			
180	mL/min	42.6			
240		49.2			
300		54.5			
1020		91.2			
1080		93.6			

Table A2.3: The raw data for the Figure 5.2: As(V) adsorption isotherms on MAS, IRAN and V-NMDG. a: Langmuir isotherm; b: Freundlich isotherm. 50 mL of solution with different adsorbents was shaking at a rate of 150 rpm for 24 hours.

MAS	MAS (g)	pH	As(V) (mg/L)	As(V) adsorbed (mg)	As(V) q (mg/g)
Intial solution	0.504	7.5	0.28		
	0.5032	7.5	1.31		
	0.5001	7.5	3.19		
	0.5006	7.5	7.53		
	0.5061	7.5	16.2		
	0.5042	7.5	32.0		
Solution after desorp-tion			0.03	0.01	0.02
			0.17	0.05	0.10
			0.91	0.10	0.21
			3.72	0.17	0.34
			10.1	0.28	0.55
			26.0	0.27	0.53
V-NMDG	**NMDG (g)**	**pH**	**As(V) (mg/L)**	**As(V) adsorbed (mg)**	**As(V) q (mg/g)**
Intial solution	0.0254	5.70	0.01		
	0.0251	5.78	0.05		
	0.026	5.84	0.10		
	0.0255	5.87	0.30		
	0.0255	5.90	0.61		
	0.025	5.95	1.30		
	0.0248	6.59	6.24		
	0.0248	6.56	11.5		
	0.0248	6.56	61.2		
	0.0253	6.56	100		
	0.0247	6.48	254		

Solution after desorp-tion		3.45	0.0001	0.001	0.02
		3.41	0.0001	0.002	0.10
		3.39	0.0002	0.005	0.18
		3.39	0.0003	0.01	0.53
		3.39	0.0007	0.03	1.08
		3.40	0.0015	0.06	2.34
		3.45	0.10	0.28	11.1
		3.51	2.07	0.42	17.1
		4.06	35.3	1.16	46.8
		4.75	64.6	1.59	63.0
		6.09	214.6	1.77	71.7
	IRAN (g)	**pH**	**As(V) (mg/L)**	**As(V) adsorbed (mg)**	**As(V) q (mg/g)**
IRAN Intial solution	0.52	7.5	0.51		
	0.528	7.5	1.02		
	0.52	7.5	50.8		
	0.518	7.5	102		
	0.525	7.5	254		
	0.503	7.5	772		
	0.501	7.5	1510		
Solution after desorp-tion Solution after desorp-tion			0.00	0.03	0.05
			0.00	0.05	0.10
			0.02	2.54	4.88
			0.05	5.08	9.80
			10.2	12.2	23.2
			373	19.9	39.6
			1079	21.6	43.1

Table A2.4: The raw data from the Figure 5.3: The adsorption of AsO_4^{3-} and PO_4^{3-} in one cartridge (2.5 g MAS). Passing through 1000 mL of solution containing 150 µg/L As(V) and 2500 µg/L P, where As(III)/As(V): ~1:4, pH: 7.0, flow rate: 50 mL/min.

Volume(mL)	As (µg/L)	P (µg/L)
20	82.97	79.9
110	87.31	32.6
210	87.22	24.5
310	86.63	19.6
410	87.57	14.8
510	87.41	19.6
610	86.55	37.5
710	85.73	24.5
910	86.53	26.1
980	85.69	31.0
In the total effluent	83.85	32.6

Table A2.5: The raw data for the Figure 6.2: Fraction of dissolved asenite after addition of $Ba(NO_3)_2$ in a solution with and without sulfate.

Batch of samples	Initial As(III) (mg/L)	Initial SO_4^{2-} (mg/L)	As(III) in solution after reaction (%)	SO_4^{2-} in solution after reaction (%)
pure As(III)				
As-Ba-11	63.5	0	96.2	BDL
As-Ba-22	66.1	0	94.1	BDL
As-Ba-33	65.3	0	94.4	BDL
As-Ba-44	65.5	0	93.3	BDL
As-Ba-55	64.4	0	91.9	BDL
As(III) +SO_4^{2-}				
As-Ba-SO4-11	33.5	1289	-	BDL
As-Ba-SO4-22	33.4	1283	60.3	BDL
As-Ba-SO4-33	33.1	1297	65.2	BDL
As-Ba-SO4-44	34.1	1521	63.9	BDL
As-Ba-SO4-55	33.9	1279	64.8	BDL

Table A2.6: The raw data for the Figure 6.3: The concentration of AsO_3^{3-} and other anions in solution after separation following the scheme in Figure 6.3.

Sample	unit	AWG-1	AWG-2	AWG-3
As(III)	µg/L	601	603	10.2
Cl^-	mg/L	55.1	52.7	0.4
NO_3^-	mg/L	0.5	554	649
SO_4^{2-}	mg/L	159	1.76	1.74
HCO_3^-	mg/L	378	0.00	0

Table A2.7: The raw data for the Figure 6.6: As(III) adsorption isotherms on MAS, IRAN and V-NMDG. a: Langmuir isotherm; b: Freundlich isotherm. 50 mL of solution with different adsorbents was shaking at a rate of 150 rpm for 24 hours.

BYC	BYC (g)	pH	As(III) (mg/L)	As(III) adsorbed (mg)	As(III) q (mg/g)
Orginal solution	0.0258	9.96	0.01		
	0.0259	9.96	0.02		
	0.0256	9.97	0.05		
	0.0258	9.94	0.31		
	0.0255	9.85	0.66		
	0.0253	9.82	1.27		
	0.0253	9.85	5.32		
	0.0253	9.86	10.7		
	0.0259	9.83	51.0		
	0.0252	9.91	97.1		
	0.0258	10.03	243		
Solutiom after desorption		9.62	0.001	0.001	0.02
		9.64	0.01	0.001	0.04
		9.57	0.01	0.002	0.08
		9.57	0.06	0.012	0.48
		9.42	0.07	0.030	1.16
		9.19	0.07	0.060	2.36
		9.21	1.07	0.212	8.39
		9.06	2.40	0.417	16.5
		9.06	31.5	0.973	37.6
		9.58	80.7	0.823	32.6
		9.89	226	0.883	34.2
MHZR	**MHZR (g)**	**pH**	**As(III)(mg/L)**	**As(III) adsorbed (mg)**	**As(III) q (mg/g)**

	0.1035	9.96	0.013		
	0.1064	9.96	0.024		
	0.1081	9.97	0.051		
	0.1059	9.94	0.305		
	0.1019	9.85	0.659		
Orginal solution	0.1018	9.82	1.27		
	0.1054	10.03	5.32		
	0.1068	9.86	11.0		
	0.1016	9.83	50.7		
	0.1035	9.91	97.1		
	0.1037	10.03	243		
		8.63	0.001	0.001	0.01
		9.07	0.001	0.001	0.01
		9.05	0.002	0.002	0.02
		9.13	0.004	0.015	0.14
Solutiom after desorption		8.63	0.006	0.033	0.32
		8.16	0.008	0.063	0.62
		9.74	0.59	0.237	2.25
		9.39	1.25	0.487	4.56
		9.89	26.7	1.202	11.8
		9.97	70.3	1.343	13.0
		9.99	218	1.283	12.4
NFO	**NFO (g)**	**pH**	**As(III)(mg/L)**	**As(III) ad-sorbed (mg)**	**As(III) q (mg/g)**
	0.0107	10.09	0.01		
	0.0107	10.05	0.03		
	0.0103	10.13	0.07		
	0.0104	10.13	0.36		
	0.0106	10.11	0.76		
Orginal solution	0.0100	10.08	1.35		
	0.0101	9.47	6.08		
	0.0109	9.60	11.0		
	0.0112	9.80	58.4		
	0.0105	9.96	130		
	0.0109	10.00	242		
		10.01	0.002	0.001	0.05
Solutiom after desorption		9.90	0.007	0.001	0.11
		9.93	0.010	0.003	0.28
		9.98	0.056	0.015	1.46

	9.99	0.17	0.029	2.77
	9.95	0.48	0.043	4.31
	9.12	4.02	0.103	10.2
	9.51	8.98	0.101	9.28
	9.83	54.7	0.184	16.4
	9.96	127	0.169	16.1
	9.98	238	0.183	16.8

Table A2.8: The raw data for Table 7.2: precipitations Ag_3AsO_4 without NH4OH. Experimental conditions: 20 mL batches with pure As(V); concentration of the added $AgNO_3$ solution: 10 g/L ; reaction time: 5 min.

Sample	$Na_2HAsO_4 \cdot 7H_2O$ (g)	Filter (g)	Ag_3AsO_4+F (g)	Ag_3AsO_4 (g)	Recovery (%)
As0707-1	0.3090	0.0921	0.3728	0.2807	61.3
As0807_1	0.1143	0.0913	0.2041	0.1128	66.6
As0807_2	0.1173	0.0914	0.2006	0.1092	62.8
As0807_3	0.1171	0.0914	0.1995	0.1081	62.3
As0807_4	0.1187	0.0916	0.1995	0.1079	61.3
As0807_5	0.1183	0.0916	0.1894	0.0978	55.8

Table A2.9: The replication of precipitation tests of Ag_3AsO_4 in different conditions.

Sample	$Na_2HAsO_4 \cdot 7H_2O$ (g)	Description	Filter (g)	Filter+Ag_3AsO_4 (g)	Ag_3AsO_4 (g)	Recovery (%)
A-1	0.0204	Without pH adjust	0.0877	0.106	0.0183	61
A-2	0.0203		0.0878	0.1088	0.0210	70
A-3	0.0211		0.0888	0.1105	0.0217	69
A-4	0.0208		0.0880	0.1101	0.0221	72
A-5	0.0211		0.0887	0.1096	0.0209	67
AH-1	0.0206	pH adjusted to7.5	0.0892	0.1228	0.0336	110
AH-2	0.0203		0.0890	0.1217	0.0327	109
AH-3	0.0204		0.0890	0.1213	0.0323	107
AH-4	0.0206		0.0888	0.1217	0.0329	108
AH-5	0.0203		0.0885	0.1211	0.0326	108
AH-6	0.0206	pH adjusted to7.5, in glove box	0.0918	0.1302	0.0384	126
AH-7	0.0199		0.0898	0.1341	0.0443	150
AH-8	0.0215		0.0900	0.1357	0.0457	143
AH-9	0.0221		0.0899	0.126	0.0361	110
AH-10	0.0201		0.0883	0.1281	0.0398	134

Table A2.10: Precipitations by using different $Na_2HAsO_4 \cdot 7H_2O$. Experimental conditions: 20 mL batches with pure As(V); concentration of the added $AgNO_3$ solution: 10 g/L ; reaction time: 5 min; pH adjusted to 7.5 with NH_4OH.

Sample	$Na_2HAsO_4 \cdot 7H_2O$ (g)	Description	Filter (g)	Ag_3AsO_4+ F (g)	Ag_3AsO_4 (g)	Recovery (%)
As1	0.0205	Original solid	0.0896	0.1192	0.0296	97.4
As2	0.0208		0.0905	0.1212	0.0307	99.6
As3	0.0209	After dried at 50 °C for 4 days	0.0805	0.1114	0.0309	99.7
As4	0.0208		0.0907	0.1201	0.0294	95.3

Table A2.11: Comparison of precipitation of Ag_3AsO_4 in different initial solution. Experimental conditions: 20 mL batches with pure As(III) and As(V); concentration of the added $AgNO_3$ solution: 10 g/L ; reaction time: 5 min; pH adjusted to 7.5 with NH_4OH or HNO_3.

Sample	$Na_2HAsO_4 \cdot 7H_2O$ (g)	Filter (g)	Solution	pH (before adjusted)	pH (after adjusted)	Filter + Ag_3AsO_4 (g)	Recovery (%)
As1	0.0202	0.0888	water	6.5	7.00	0.105	54.1
As2	0.0269	0.0894	water	6.5	7.60	0.1087	48.4
NaAs1	0.0212	0.0907	0.5 M NaOH	6.94	7.00	0.1205	94.8
NaAs2	0.0239	0.0902	0.5 M NaOH	6.4	7.50	0.1245	96.8

Table A2.12: Test for precipitation of arsenite in pure water. Experimental conditions: 20 mL batches with pure As(III); concentration of the added $AgNO_3$ solution: 10 g/L ; reaction time: 5 min; pH adjusted to 7.5 with NH_4OH.

Sample	As(III) (mg/L)	Filter (g)	Filter + Ag_3AsO_3 (g)	Ag_3AsO_3 (g)	Precipitation (%)
As3-0 ($NaAsO_2$)	solid				
As3-1	1	0.0874	0.1152	0.0278	93.0
As3-2	1	0.0881	0.1166	0.0285	95.4
As3-3	1	0.0880	0.1158	0.0278	93.0
As3-4	1	0.0887	0.1168	0.0281	94.0
As3-5	1	0.0874	0.116	0.0286	95.7

Table A2.13: The raw data fro table7.3: the affection of different concentration of As(V) to precipitation in pure water. 20 m of solution with different concentration of As(V) was precipitated with $AgNO_3$ in excess at a pH range of 7.3-7.5 in 10 minutes.

Sample	As(V) (mg/L)		Precipitation (%)	10 g/L $AgNO_3$ (mL)
	Before precipitation	After precipitation		
AsV-0.5	0.6	0.005	99.3	0.5
AsV-1.0	1.3	0.002	99.9	0.5
AsV-5.0	5.1	0.015	99.7	0.5
AsV-10	10.2	0.004	100.0	0.5
AsV-50	50.8	0.024	100.0	1
AsV-100	102	0.131	99.9	2
AsV-250	254	0.208	99.9	4

Table A2.14: The raw data for Table7.3: the affection of different concentration of As(III)) to precipitation in pure water. 20 m of solution with different concentration of As(III) was precipitated with $AgNO_3$ in excess at a pH range of 7.3-7.5 in 10 minutes.

Sample	AsIII)(mg/L)		Precipitation (%)	10 g/L $AgNO_3$ (mL)
	Before precipitation	After precipitation		
AsIII-0.5	0.68425	0.607	11.3	0.5
AsIII-1	1.4	1.246	8.9	0.5
AsIII-5	6.8	5.815	15.0	0.5
AsIII-10	13.7	11.729	14.3	0.5
AsIII-50	54.7	11.949	78.2	1
AsIII-100	109.5	0.198	99.8	2
AsIII-250	273.7	1.844	99.3	4
AsIII-500	547.4	8.748	98.4	8

Table A2.15: The raw data for Figure 7.4: Variation with time of the oxygen isotope composition of dissolved AsO_4^{3-} ($[AsO_4^{3-}]$: 450 mg/L) for pH ranging from 1 to 10 under 25 °C and 70 °C. The initial $\delta^{18}O$ value of H_2O was -12.0 ‰.

pH range	Time (h)	$\delta^{18}O$ value (AsO_4^{3-}) ‰VSMOW (25 °C)				$\delta^{18}O$ value (AsO_4^{3-}) ‰VSMOW (70 °C)		
		Average	Stdev	n		Average	Stdev	n
pH 1	0	1.51	0.17	4		1.51	0.17	4
	3	0.14	0.02	4		-3.13	0.08	4
	6	-0.35	0.10	4		-3.12	0.08	4
	12	-0.44	0.41	4		-3.08	0.23	4
	24	-0.28	0.29	4		-3.40	0.23	4
	48	-0.35	0.38	4		-2.99	0.15	4
	96	0.05	0.06	4		-3.18	0.02	4
	168	0.15	0.19	4		-3.09	0.03	4
	312	-0.45	0.21	4		-3.21	0.16	4
	648	-1.00	0.02	3	552 (h)	-3.25	-	1
pH 5	0	1.51	0.17	4		1.51	0.17	4
	3	-3.13	0.08	4		-6.98	0.09	4
	6	-3.12	0.08	4		-7.00	0.02	4
	12	-3.08	0.23	4		-7.10	0.06	4
	24	-3.40	0.23	4		-7.01	0.01	4
	48	-2.99	0.15	4		-7.22	0.28	4
	96	-3.18	0.02	4		-5.96	0.00	4
	168	-3.09	0.03	4		-7.05	0.05	4
	312	-3.21	0.16	4		-7.58	0.02	4
	648	-3.86	0.14	3	552 (h)	-7.41	-	1
pH 10	0	1.51	0.17	4		1.51	0.17	2
	3	1.17	0.03	4		-11.0	0.32	2
	6	0.34	0.16	4		-11.3	0.0	2
	12	-0.19	0.09	4		-11.4	0.0	2
	24	-1.65	0.13	4		-10.9	0.0	2
	48	-3.63	0.02	4		-11.3	0.2	2
	96	-5.46	0.09	4		-11.0	0.2	2
	168	-6.88	0.15	4		-11.2	0.0	2
	312	-7.27	0.19	4		-11.4	0.0	2
	648	-7.86	0.04	2	552 (h)	-	-	-

In der Reihe „Karlsruher Geochemische Hefte" (ISSN 0943-8599) sind erschienen:

Band 1: U. Kramar (1993)
Methoden zur Interpretation von Daten der geochemischen Bachsedimentprospektion am Beispiel der Sierra de San Carlos/ Tamaulipas Mexiko

Band 2: Z. Berner (1993)
S-Isotopengeochemie in der KTB - Vorbohrung und Beziehungen zu den Spuren-elementmustern der Pyrite

Band 3: J.-D. Eckhardt (1993)
Geochemische Untersuchungen an jungen Sedimenten von der Galapagos-Mikroplatte: Hydrothermale und stratigraphisch signifikante Signale

Band 4: B. Bergfeldt (1994)
Lösungs- und Austauschprozesse in der ungesättigten Bodenwasserzone und Auswirkungen auf das Grundwasser

Band 5: M. Hodel (1994)
Untersuchungen zur Festlegung und Mobilisierung der Elemente As, Cd, Ni und Pb an ausgewählten Festphasen unter besonderer Berücksichtigung des Einflusses von Huminstoffen.

Band 6: T. Bergfeldt (1995)
Untersuchungen der Arsen- und Schwermetallmobilität in Bergbauhalden und kontaminierten Böden im Gebiet des Mittleren Schwarzwaldes

Band 7: M. Manz (1995)
Umweltbelastungen durch Arsen und Schwermetalle in Böden, Halden, Pflanzen und Schlacken ehemaliger Bergbaugebiete des Mittleren und Südlichen Schwarzwaldes.

Band 8: J. Ritter (1995)
Genese der Mineralisation Herrmanngang im Albtalgranit (SE-Schwarzwald) und Wechselwirkungen mit dem Nebengestein.

Band 9: J. Castro (1995)
Umweltauswirkungen des Bergbaus im semiariden Gebiet von Santa Maria de la Paz, Mexiko.

Band 10: T. Rüde (1996)
Beiträge zur Geochemie des Arsens.

Band 11: J. Schäfer (1998)
Einträge und Kontaminationspfade Kfz-emittierter Platin-Gruppen-Elemente (PGE) in verschiedenen Umweltkompartimenten.

Band 12: M. A. Leosson (1999)
A Contribution to the Sulphur Isotope Geochemistry of the Upper Continental Crust: The KTB Main Hole - A Case Study

Band 13: B. A. Kappes (2000)
Mobilisierbarkeit von Schwermetallen und Arsen durch saure Grubenabwässer in Bergbaualtlasten der Ag-Pb-Zn-Lagerstätte in Wiesloch

Band 14: H. Philipp (2000)
The behaviour of platinum-group elements in petrogenetic processes: A case study from the seaward-dipping reflector sequence (SDRS), Southeast Greenland volcanic rifted margin

Band 15: E. Walpersdorf (2000)
Nähr- und Spurenelementdynamik im Sediment/Wasser-Kontaktbereich nach einer Seekreideaufspülung – Pilotstudie Arendsee –

Band 16: R. H. Kölbl (2000)
Models of hydrothermal plumes by submarine diffuse venting in a coastal area: A case study for Milos, South Aegean Volcanic Arc, Greece

Band 17: U. Heiser (2000)
Calcium-rich Rhodochrosite layers in Sediments of the Gotland Deep, Baltic Sea, as Indicators for Seawater Inflow

In der Fortsetzungsreihe „Karlsruher Mineralogische und Geochemische Hefte" (ISSN 1618-2677) sind bisher erschienen:

Band 18: S. Norra (2001)
Umweltgeochemische Signale urbaner Systeme am Beispiel von Böden, Pflanzen, und Stäuben in Karlsruhe

Band 19: M. von Wagner, Alexander Salichow, Doris Stüben (2001)
Geochemische Reinigung kleiner Fließgewässer mit Mangankiesen, einem Abfallsprodukt aus Wasserwerken (GReiFMan)

Band 20: U. Berg (2002)
Die Kalzitapplikation als Restaurierungsmaßnahme für eutrophe Seen – ihre Optimierung und Bewertung

Band 21: Ch. Menzel (2002)
Bestimmung und Verteilung aquatischer PGE-Spezies in urbanen Systemen

Band 22: P. Graf (2002)
Meta-Kaolinit als latent-hydraulische Komponente in Kalkmörtel

Band 23: D. Buqezi-Ahmeti (2003)
Die Fluidgehalte der Mantel-Xenolithe des Alkaligesteins-Komplexes der Persani-Berge, Ostkarpaten, Rumänien: Untersuchungen an Fluid-Einschlüssen

Band 24: B. Scheibner (2003)
Das geochemische Verhalten der Platingruppenelemente bei der Entstehung und Differenzierung der Magmen der Kerguelen-Flutbasaltprovinz (Indischer Ozean)

Ab Band 25 erscheinen die Karlsruher Mineralogischen und Geochemischen Hefte bei KIT Scientific Publishing online unter der Internetadresse http://www.ksp.kit.edu
Auf Wunsch sind bei KIT Scientific Publishing gedruckte Exemplare erhältlich („print on demand").

Band 25: A. Stögbauer (2005)
Schwefelisotopenfraktionierung in abwasserbelasteten Sedimenten - Biogeochemische Umsetzungen und deren Auswirkung auf den Schwermetallhaushalt

Band 26: X. Xie (2005)
Assessment of an ultramicroelectrode array (UMEA) sensor for the determination of trace concentrations of heavy metals in water

Band 27: F. Friedrich (2005)
Spectroscopic investigations of delaminated and intercalated phyllosilicates

Band 28: L. Niemann (2005)
Die Reaktionskinetik des Gipsabbindens: Makroskopische Reaktionsraten und Mechanismen in molekularem Maßstab

Band 29: F. Wagner (2005)
Prozessverständnis einer Naturkatastrophe: eine geo- und hydrochemische Untersuchung der regionalen Arsen-Anreicherung im Grundwasser West-Bengalens (Indien)

Band 30: F. Pujol (2005)
Chemostratigraphy of some key European Frasnian-Famennian boundary sections (Germany, Poland, France)

Band 31: Y. Dikikh (2006)
Adsorption und Mobilisierung wasserlöslicher Kfz-emittierter Platingruppenelemente (Pt, Pd, Rh) an verschiedenen bodentypischen Mineralen

Band 32: F. I. Müller (2007)
Influence of cellulose ethers on the kinetics of early Portland cement hydration

Band 33: H. Haile Tolera (2007)
Suitability of local materials to purify Akaki Sub-Basin water

Band 34: M. Memon (2008)
Role of Fe-oxides for predicting phosphorus sorption in calcareous soils

Band 35: S. Bleeck-Schmidt (2008)
Geochemisch-mineralogische Hochwassersignale in Auensedimenten und deren Relevanz für die Rekonstruktion von Hochwasserereignissen

Band 36: A. Steudel (2009)
Selection strategy and modification of layer silicates for technical applications

Band 37: E. Eiche (2009)
Arsenic mobilization processes in the red river delta, Vietnam: Towards a better understanding of the patchy distribution of dissolved arsenic in alluvial deposits

Band 38: N. Schleicher (2012)
Chemical, Physical and Mineralogical Properties of Atmospheric Particulate Matter in the Megacity Beijing

Band 39: H. Neidhardt (2012)
Arsenic in groundwater of West Bengal

Band 40: X. Tang (2014)
Separating arsenic oxyanions from natural waters for oxygen isotope analysis